Heat Transfer

Lessons with Examples Solved by Matlab

FIRST EDITION

By Tien-Mo Shih

Department of Physics, Xiamen University
ICAM, University of California, Davis

❀ cognella™
San Diego, CA

Bassim Hamadeh, CEO and Publisher
Christopher Foster, Vice President
Michael Simpson, Vice President of Acquisitions
Jessica Knott, Managing Editor
Jess Busch, Creative Director
Kevin Fahey, Marketing Program Manager
Melissa Barcomb, Acquisitions Editor
Sarah Wheeler, Senior Project Editor
Erin Escobar, Licensing Associate

First published in the United States of America in 2012 by Cognella, Inc.

Trademark Notice: Product or corporate names may be trademarks or registered trademarks, and are used only for identification and explanation without intent to infringe.

15 14 13 12 11   1 2 3 4 5

Printed in the United States of America

ISBN: 978-1-60927-544-0

✿ cognella™

www.cognella.com   800.200.3908

Dedication

To my wife, Tingting Zhu, whose love and vegetarian cooking are the primary driving force of this writing endeavor.

Contents

Preface   xix

Lesson 1: Introduction (What and Why)   1

1. Heat Transfer Is an Old Subject   1
   1-1 Partially True   2
   1-2 Modifications of the Perception   2

2. What Is the Subject of Heat Transfer?   3
   2-1 Definition   3
   2-2 Second Income   3

3. Candle Burning for Your Birthday Party   5
   3-1 Unknowns and Governing Equations   5
   3-2 Solution Procedure   6

4. Why Is the Subject of Heat Transfer Important?   7
   4-1. Examples of Heat Transfer Problems in Daily Life   7
   4-2 Examples of Heat Transfer Problems in Industrial Applications   8

5. Three Modes of Heat Transfer   8

6. Prerequisites   9

7. Structure of the Textbook   9

8. Summary   10

9. References   10

10. Exercise Problems   11

11. Appendix   11
    A-1 Sum Up 1, 2, 3,…, 100   11
    A-2 Plot f(x)   12
    A-3 Solve Three Linear Equations   12

Lesson 2: Introduction (Three Laws)   13

1. Fourier's Law   13
2. Law of Convective Heat Transfer   15
3. Wind-Chill Factor (WCF)   16
4. Stefan-Boltzmann Law of Radiative Emission   17
5. Sheet Energy Balance   18
6. Formation of Ice Layers on Car Windshield and Windows   18
    6-1 Clear Sky Overnight   18
    6-2 Cloudy Sky Overnight   20
7. Rule of Assume, Draw, and Write (ADW)   20
8. Summary   22
9. References   22
10. Exercise Problems   22
11. Appendix: Taylor's Series Expansion   23

Lesson 3: One-Dimensional Steady State Heat Conduction   25

1. Governing Equation for T(x) or T(i)   25
2. A Single-Slab System   26
3. A Two-Slab System   28
4. Three or More Slabs   29
5. Severe Restrictions Imposed by Using Electrical Circuit Analogy   30
6. Other Types of Boundary Conditions   30
7. Thermal Properties of Common Materials — Table 1   31
8. Summary   31
9. References   31
10. Exercise Problems   31
11. Appendix: A Matlab Code for One-D Steady-State Conduction   32

Lesson 4: One-Dimensional Slabs with Heat Generation 35

1. Introduction   35

2. Governing Equations   36

3. Heat Conduction Related to Our Bodies   37
    3-1 Estimate the Heat Generation of Our Bodies   37
    3-2 A Coarse Grid System   39
    3-3 Optimization   40

4. Discussions   41
    4-1 Radiation Boundary Condition   41
    4-2 Sketch the Trends and Check our Speculations
        with Running the Matlab Code   42
    4-3 Check the Global Energy Balance   43

5. Summary   43
6. References   43
7. Exercise Problems   43
8. Appendix: A Matlab Code for Optimization   44

Lesson 5: One-Dimensional Steady-State Fins 47

1. Introduction   48
    1-1 Main Purpose of Fins   48
    1-2 Difference between 1-D Slabs and 1-D Fins   49
    1-3 A Quick Estimate   49

2. Formal Analyses   51
    2-1 Governing Equation   51
    2-2 A Standard Matlab Code Solving $T(x)$ and q_b for a Fin   51
    2-3 Equivalent Governing Differential Equation   52

3. Fins Losing Radiation to Clear Sky Overnight   52
4. A Seemingly Puzzling Phenomenon   54
5. Fin Efficiency   55
6. Optimization   55
    6-1 Constraint of Fixed Total Volume   56
    6-2 Optimization for Rod Bundles   56

7. Summary   58

8. References   58

9. Exercise Problems   58

10. Appendix   59

    A-1 Explanation of the Seemingly Puzzling Phenomenon   59

    A-2 Cubic Temperature Profile   60

    A-3 Constraint of Fixed Fin Mass (or Volume)   62

    A-4 Fin Bundles and Optimization   63

Lesson 6: Two-Dimensional Steady-State Conduction   65

1. Governing Equations   66

    1-1 Derivation of the General Governing Equation   66

    1-2 Three-Interior-Node System   67

    1-3 A Special Case   70

2. A Standard Matlab Code Solving 2-D Steady-State Problems   70

3. Maximum Heat Loss from a Cylinder Surrounded by Insulation Materials   70

    3-1. Governing Equation   71

    3-2 An Optimization Problem   72

4. Summary   72

5. References   73

6. Exercise Problems   73

7. Appendix   74

    A-1 Gauss-Seidel Method   74

    A-2 A Standard 2-D Steady State Code   75

    A-3 Optimization Problem in the Cylindrical Coordinates   77

Lesson 7: Lumped-Capacitance Models (Zero-Dimension Transient Conduction)   79

1. Introduction   79

    1-1 Adjectives   79

    1-2 Justifications   80

1-3 Examples ... 80

1-4 Objectives and Control Volumes ... 81

1-5 The Most Important Term and the First Law of Thermodynamics ... 81

2. Detailed Analyses of a Can-of-Coke Problem ... 82
   2-1 Modeling ... 82

2-2 Assumptions in the Modeling ... 83

2-3 A Matlab Code Computing T(t) for the Can of Coke Problem ... 83

2-4 Discussions ... 84

3. When Is It Appropriate to Use the Lumped-Capacitance Model? ... 85
   3-1 A Three-Node Example ... 85

3-2 An Analogy ... 86

4. A Simple Way to Relax the Bi < 0.002 Constraint ... 87

5. Why Stirring the Food When We Fry It? ... 88

6. Summary ... 90

7. References ... 90

8. Exercise Problems ... 90

9. Appendix ... 91
   A-1 Discussions of Adjectives ... 91

A-2 Comparisons of Three Cases ... 92

Lesson 8: One-Dimensional Transient Heat Conduction ... 95

1. Kitchen Is a Good Place to Learn Heat Transfer ... 95
   1-1 Governing Equation of One-D Transient Heat Conduction ... 95

1-2 The Generic Core of One-D Transient Code ... 97

1-3 Various Boundary Conditions ... 97

1-4 A One-D Transient Matlab Code ... 97

1-5 Global Energy Balance ... 99

2. Other One-D Transient Heat Conduction Applications ... 100
   2-1. Semi-Infinite Solids ... 100

2-2. Revisit Fin Problems ... 100

2-3. Multi-layer Slabs with Heat Generation ... 102

3. Differential Governing Equation for One-D Transient Heat Conduction   103

4. Summary   103

5. References   103

6. Exercise Problems   103

7. Appendix   104
    A-1 Semi-infinite Solids   104

    A-2 Transient 1-D Fins   105

    A-3 Example 4-1 Revisited   106

Lesson 9: Two-D Transient Heat Conduction   109

1. Governing Equation for T(i, j)   109
    1-1 General Case   109

    1-2 Special Cases   110

2. A Standard Matlab Code for Readers to Modify   113

3. Speculation on Steel Melting in Concrete Columns during 9/11   113

4. Possible Numerical Answers   114

5. Use a Two-D code to Solve One-D Transient Heat Conduction Problems   115

6. Exact Solutions for Validation of Codes   116

7. Advanced Heat Conduction Problems   117
    7-1 Moving Interface (or called Stefan Problem)   117

    7-2. Irregular Geometries   118

    7-3. Non-Fourier Law (Hyperbolic-Type Heat Conduction Equation)   118

8. Summary   120

9. References   120

10. Exercise Problems   120

11. Appendix   121
    A-1 A Standard Transient 2-D Code for Readers to Modify   121

    A-2. Investigation of Transient 1-D Heat Conduction for a Steel Column   122

    A-3. Investigation of Transient 2-D Heat Conduction for the Concrete Column   123

Lesson 10: Forced-Convection External Flows (I)   125

1. Soup-Blowing Problem   125

2. Boundary- Layer Flows   128

3. A Cubic Velocity Profile   130
   3–1 Determine the coefficients   130

   3-2 Application of Eq. (3)   131

4. Summary   133

5. References   133

6. Exercise Problems   133

7. Appendix: Momentum Balance over an Integral Segment   136

Lesson 11: Forced-Convection External Flows (II)   139

1. Nondimensionalization (abbreviated as Ndm)   139

2. Important Dimensionless Parameters in Heat Transfer   144

3. Derivation of Governing Equations   144

4. Categorization   145

5. Summary   146

6. References   146

7. Exercise Problems   146

8. Appendix: Steady-State Governing Equations   148
   A-1 Derivation of Governing Equation for u   148

   A-2 Governing Equation for T   152

1. Preliminary   153

2. A Classical Approach Reported in the Literature   158

3. Steps to Find Heat Flux at the Wall (from the Similarity Solution)   156

4. The Nu Correlation and Some Discussions   157

5. Derivation of Nu = G (Re, Pr) by Ndm   159

6. Finding Eq. (6b) by Using a Quick and Approximate Method   160

   6-1. A Quick and Approximate Method   160

   6-2. A Matlab code generating the Nu correlation   161

7. An Example Regarding Convection and Radiation Combined   163

8. Possible Shortcomings of Nu Correlations   164

9. Brief Examination of Two More External Flows   164

10. Summary   165

11. References   165

12. Exercise Problems   166

13. Appendix [to find $f'(\eta)$ and the value of $f''(0)$]   168

   A-1 A Matlab Code for Solving the Blasius Similarity Equation   168

   A-2 Table of η, f, f', f'' Distributions   169

   A-3 Brief Explanations of the Table   171

Lessons 13: Internal Flows (I)—Hydrodynamic Aspect   173

1. Main Differences Between External Flows and Internal Flows   174

2. Two Regimes (or Regions)   174

3. A Coarse Grid to Find u, v, and p in the Developing Regime   176

4. An Analytical Procedure of Finding u(y) in the Fully Developed Regime 178

5. Application of the Results   182

6. Which Value Should We Use?   183

7. Ndm and Parameter Dependence   184

8. Summary   184

9. References   184

10. Exercise Problems   185

11. Appendix: A Matlab Code for Finding u, v, and p in the Developing Regime   187

Lessons 14: Internal Flows (II)—Thermal Aspect   189

1. Definition of Tm   189

2. Definition of Thermally Fully Developed Flows   190

3. Justification of $\partial T/\partial x$ = constant   191

4. A Beneficial Logical Exercise of Genetics   193

5. Summary   194

6. References   194

7. Exercise Problems   194

Lessons 15: Internal Flows (III)—Thermal Aspect   197

1. Derivation of Nu Value for Uniform q''s   197

2. Important Implications of Eq. (5)   199

3. Derivation of Nu value for uniform Ts   200

    3-1 Justification   201

    3-2 Solution Procedure   202

4. Let the Faucet Drip Slowly   204

5. Summary   207

6. References   207

7. Exercise Problems   207

8. Appendix: A Matlab Code Computing Nu for the Case of Uniform Ts   208

Lesson 16: Free Convection   211

1. Definition of Free Convection   211

2. Definition of Buoyancy Force   212

    2-1 Buoyancy Force on an Object   212

    2-2 Buoyancy Force on a Control Volume in the Flow   213

    2-3 Density Variations   214

3. The Main Difference Between Free Convection and Forced Convection   214

4. How Does Gr Number Arise?   215

5. δ_T and δ in Free Convection   216

6. A Four-Cell Buoyancy-Driven Flow in a Square Enclosure   217

    6-1 Description   217

    6-2 Derivation of Governing Equations   218

    6-3 Discussions of the Result   219

7. Does Lighting a Fire in Fireplace Gain Net Energy for the House?   219

    7-1 Governing Equations   220

    7-2 Nomenclature in the Code   220

    7-3 Matlab Code   220

8. Solar Radiation-Ice Turbine   221

    8-1 Description of the Machine   221

    8-2 Some Analyses   222

9. Free Convection over a Vertical Plate   223

10. Summary   224

11. References   224

12. Exercise Problems   224

13. Appendix: A Matlab Code Computing Buoyancy-Driven Flows in Four-Cell Enclosures 224

Lesson 17: Turbulent Heat Convection   227

1. Introduction   228
   1-1 Speed of Typical Flows   228

   1-2 Frequencies of Turbulence and Molecular Collision   228

   1-3 Superposition   228

2. A Fundamental Analysis   229
   2-1 Governing Equations   229

   2-2 Zero-Equation Turbulence Model   230

   2-3 Discretized Governing Equation   231

3. Matlab Codes   232
   3-1 Laminar Flows in Two-Parallel-Plate Channels   232

   3-2 Turbulent Flows   233

   3-3 Laminar Flows in Circular Tubes   233

4. Dimples on Golf Balls   234

5. Summary   235

6. References   235

7. Exercise Problems   235

8. Appendix   235
   A-1 Laminar Flows in Circular Tubes   235

   A-2 Fully Developed Turbulent Flows in Planar Channels   236

Lesson 18: Heat Exchangers and Other Heat Transfer Applications   239

1. Types of Heat Exchangers   239
2. A Fundamental Analysis   240
3. A Traditional Method to Find Heat Exchange   241
4. A Matlab Code   242
5. Comments on the Code   244
6. Other Applications in Heat Transfer   244

    6-1 Combustion and Low-Temperature Chemical Reactions   244

    6-2 Jets, Plumes, and Wakes   245

    6-3 Optimization   245

    6-4 Porous Media   245

    6-5 Radiation with Participating Gases   245

    6-6 Two-Phase Flows   245

7. Summary   246
8. References   246
9. Exercise Problems   247

Lesson 19: Radiation (I)   249

1. Fundamental Concepts   249
    1-1 Main Difference between Radiation and Convection   250

    1-2 Adjectives Used for Radiative Properties   250

2. Blackbody Radiation   251
    2-1. Definition of a Blackbody Surface   251

    2-2. Planck Spectral Distribution   252

    2-3. Stefan-Boltzmann Law   253

3. A Coffee Drinking Tip   254

4. Fractions of Blackbody Emission   256

5. Summary   257

6. References   257

7. Exercise Problems   257

8. Appendix   258
   A-1 Finding the Value of Stefan-Boltzmann Constant   258
   A-2 Table 19-1 Fractions of Blackbody Emission   259

Lesson 20: Radiation (II)   261

1. Emissivity   261

2. Three Other Radiative Properties   263

3. Solar Constant and Effective Temperature of the Sun   264

4. Gray Surfaces   266

5. Kirchhoff's Law   268

6. Energy Balance over a Typical Plate   269

7. Greenhouse Effect (or Global Warming)   269

8. Steady-State Heat Flux Supplied Externally by Us   271

9. Find Steady-State Ts Analytically   272

10. Find Steady-State Ts Numerically   272

11. Find Unsteady Ts Not Involved with the Spectral Emissivity   273

12. Find Unsteady Ts Involved with the Spectral Emissivity   274

13. Find Unsteady Ts with Parameters Being Functions of Wavelength and Time   275

14. Summary   277

15. References   277

16. Exercise Problems   277

1. View Factors (or Shape Factors, Configuration Factors)   281
    1-1 Definition of the View Factor, F_{12} 282

    1-2 Reciprocity Rule   282

    1-3 Energy Conservation Rule   282

    1-4 View Factor for a Triangle   283

2. Black Triangular Enclosures   283

3. Gray Triangular Enclosures   285
    3-1 Definition of Radiosity, J   286

4. Two Parallel Gray Plates with A1 = A2   288

5. Radiation Shield   289

6. Summary   290

7. References   290

8. Exercise Problems   290

Preface

Heat Transfer: Lessons with Examples Solved by Matlab instructs students in heat transfer, and cultivates independent and logical thinking ability. The book focuses on fundamental concepts in heat transfer and can be used in courses in Heat Transfer, Heat and Mass Transfer, and Transport Processes. It uses numerical examples and equation solving to clarify complex, abstract concepts such as Kirchhoff's Law in Radiation.

Several features characterize this textbook:

- It includes real-world examples encountered in daily life;
- Examples are mostly solved in simple Matlab codes, readily for students to run numerical experiments by cutting and pasting Matlab codes into their PCs;
- In parallel to Matlab codes, some examples are solved at only a few nodes, allowing students to understand the physics qualitatively without running Matlab codes;
- It places emphasis on "why" for engineers, not just "how" for technicians.

Adopting instructors will receive supplemental exercise problems, as well as access to a companion website where instructors and students can participate in discussion forums amongst themselves and with the author.

Heat Transfer is an ideal text for students of mechanical, chemical, and aerospace engineering. It can also be used in programs for civil and electrical engineering, and physics. Rather than simply training students to be technicians, *Heat Transfer* uses clear examples, structured exercises and application activities that train students to be engineers. The book encourages independent and logical thinking, and gives students the skills needed to master complex, technical subject matter.

Tien-Mo Shih received his Ph.D. from the University of California, Berkeley, and did his post-doctoral work at Harvard University. From 1978 until his retirement in 2011 he was an Associate Professor of Mechanical Engineering at the University of Maryland, College Park, where he taught courses in thermo-sciences and numerical methods. He remains active in research in these same areas. His book, *Numerical Heat Transfer*, was translated into Russian and Chinese, and subsequently published by both the Russian Academy of Sciences and the Chinese Academy of Sciences. He has published numerous research papers, and has been invited regularly to write survey papers for *Numerical Heat Transfer Journal* since 1980s.

Lesson 1
Introduction (What and Why)

In the very first lesson of this textbook, we will attempt to describe what the subject of heat transfer is and why it is important to us. In addition, three modes of heat transfer—namely, conduction, convection, and radiation—will be briefly explained. The structure of this textbook is outlined.

Nomenclature

c_v = specific heat, J/kg-K
Q = heat transfer, J
q = heat flow rate or heat transfer rate, W
q'' = heat flux, W/m^2
T = temperature, in C or K (only when radiation is involved, we use K).
U = internal energy of a control volume, J
V = volume of the piston-cylinder system, m^3

1. Heat Transfer Is an Old Subject

On one occasion the author overheard a casual conversation between two students:
"How did you like your heat transfer course last semester?"
"Well, it is OK, I guess; it is kind of *old*."

1-1 Partially True

The student's remark may be partially true. The subject of heat transfer is at least as old as when Nicolas Carnot designed his Carnot engine in 1824, where Q, the heat transfer in Joules, was needed for interactions between the piston-cylinder system and the reservoirs during expansion and compression isothermal processes.

Because it is an old subject, and because the course has been traditional, many textbooks include solution procedures that were used in old times before computers were invented or not yet available to the general public. A notable example is Separation of Variables, a way to solve partial differential equations. Without computers, highly ingenious mathematical theories, assumptions, and formulations must be introduced. Eigenvalues and Fourier's series are also included. Students are overwhelmed with such great amount of math. It is no wonder they feel bored.

Today, with the advent of computers and associated numerical methods, not only can these equations be solved using simple numerical methods, but also the problems no longer need to be highly idealized to fit the restrictions imposed by the Separation of Variables. In the author's opinion, this subject of Separation of Variables may not belong to textbooks that are supposed to deliver heat transfer knowledge within a short semester. At most, it should be moved to appendices, or taught in a separate math-related course.

1-2 Modifications

On the other hand, an old subject can be different from old people, whom young college students love, but may feel bored to play with. For example, history is an old subject. It can become either dull or interesting depending on the materials presented. Some possible modifications can be made for the old subject, like heat transfer, as suggested below:

(a) Useful, interesting and contemporary topics can be introduced.
(b) The modeling (that converts given problems into equations) can be conducted over control volumes of finite sizes, not of differential sizes, to avoid the appearance of differential equations.
(c) Simple numerical methods that do not require much math can be introduced and used to solve these algebraic equations.
(d) Matlab software can be further used to eliminate routine algebraic and arithmetic manipulations. In the Matlab environment, results can also be readily plotted for students to see the trends of heat transfer physics.
(e) Finally, we can try to include as many daily-life problems as possible, in substitution of some industrial-application problems. In college education, the most important objective is to get students to be motivated and interested. Once they

are, and have learned fundamentals, they will have plenty of opportunities to approach industrial-application problems after graduation.

2. What Is the Subject of Heat Transfer?

2-1 Definition

(a) It consists of three sub-topics: Conduction, Convection, and Radiation.
(b) It contains information regarding how energy is transferred from places to places or objects to objects, how much the quantity of this energy is, the methods of finding this quantity, and associated discussions.
(c) In slightly more details, we may state that the subject is for us to attempt to find temperature (T) distributions and histories in a system. We can further imagine that, as soon as we have found $T(x, y, z, t)$, then we are home. At home, we get to do things very leisurely and conveniently, such as lying on the couch, watching TV, and fetching a can of beer from the refrigerator. Similarly, after having found $T(x, y, z, t)$, we can find other quantities related to energy (or heat) transfer at our leisure and convenience. Therefore, let us keep in mind that finding $T(x, y, z, t)$ is the key in the subject of heat transfer.

2-2 Second Income

Heat transfer is also intimately related to the subject of thermodynamics. In the latter, generally, the quantity Q is either given, or calculated from the first law of thermodynamics. For example, during the isothermal expansion process from state 1 to state 2 for a piston-cylinder system containing air, which is treated as an ideal gas, we have

$$\Delta U = Q + W, \tag{1}$$

where Q and W denote heat transfer flowing ***into*** the system, and work done ***to*** the system, respectively. For an ideal gas inside a piston-cylinder system undergoing an isothermal process,

$$\Delta U = mc_v(T_2 - T_1) = 0,$$

and

$$W = -mRT \ln(V_2/V_1).$$

Therefore,

$$Q = mRT \ln(V_2/V_1). \tag{2}$$

Derivation of Eq. (2) dictates that Q be found based on the first law. In other words, the first law is already "consumed" by our desire to find Q. It cannot be re-used to find other quantities.

The subject of heat transfer is to teach us how to find Q from a different resource. The relationship between heat transfer and thermodynamics can be better understood with the following analogy.

John lives with his parents (thermodynamics). The household has been supported by only one source of income (the first law of thermodynamics), earned by his parents. John (Q) has been living under the same roof with his parents, being a good student all the way from a little cute kid in kindergarten to an adult in college. Soon it will be also his responsibility to seek an independent source of income (heat transfer), so that he can move out of his parents' house and get married.

Example 1-1

How long does it take for a can of Coke, taken out from the refrigerator at 5C, to warm up to 15C in a room at 25C? Relevant data are: m = 0.28 kg, c_v = 4180 J/kg-K, and A = $0.01m^2$.

Sol: Let us attempt to find the answer by first writing down the first law of thermo-dynamics given by Eq. (1):

$$\Delta U = Q, \tag{3a}$$

in which W vanishes because there is none. From the subject of thermodynamics, we have learned

$$\Delta U = mc_v\Delta T.$$

Therefore, Eq. (1a) can be changed to

$$mc_v\Delta T = Q. \tag{3b}$$

At this juncture, we see that there are two unknowns in Eq. (3b), T and Q. We can scratch our heads hard, and we are still unable to find another equation. In the analogy given above, the first law is the only income of the household. We need the second income, which is the second equation to solve Eq. (3b). This second equation is to be provided by the knowledge gained from the subject of heat transfer.

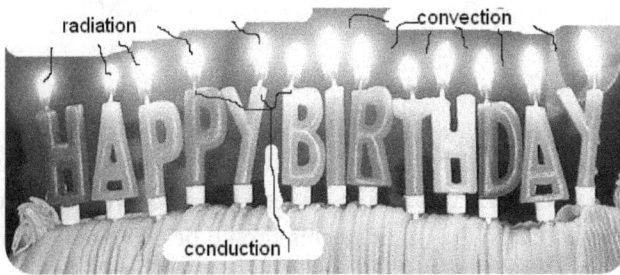

See Problems 1-1 and 1-2.

3. Candle Burning for Your Birthday Party

You are celebrating your last birthday party before you graduate with a glamorous degree. There are candles lighted on the cake for you to make a wish on and blow out. Candle fires can also serve as a good example to illustrate heat transfer, with wax melting, hot air rising, and flame illuminating associated with conduction, convection, and radiation.

3-1 Unknowns and Governing Equations

You like to buy five boxes of candles. Your parents, being frugal, like to buy only three boxes of candles. So, the first urgent question is: how long does it take for a candle of a given size to finish burning? To find out the burning rate of a candle is, by no means, a trivial question. We need to know the solution of quite a few variables in order to determine the rate. Let us count the number of variables (unknowns) and the number of governing equations below. These two numbers should exactly be equal, no more, no fewer.

(a) Three momentum equations govern the flow velocities in x, y, and z directions, u, v, and w.
(b) Conservation of energy, which is essentially the first law of thermodynamics mentioned above, governs T.
(c) Conservation of species governs mass fractions of fuel, air, and the combustion product, including Y_{fuel}, Y_{O_2}, Y_{N_2}, Y_{CO_2}, and Y_{H_2O}.
(d) Conservation of mass governs the density of unburned air, ρ
(e) The equation of state, will be responsible for $p = \rho RT$. Note that p and ρ in (d) and (e) may be switched.

In particular, in category (b), the equation takes the form of

$$\text{enthalpy\_in} + \text{conduction\_in} + \text{radiation\_in} + \text{combustion\_in} = (\Delta U/\Delta t)_{control\text{-}volume}. \quad (4)$$

The control volume can be the entire flame if a crude solution is sought; or it can be small computational cells, $\Delta x \Delta y \Delta z$, for us to seek more accurate solutions. Regardless, the essential point is that, in Eq. (4), all terms should be eventually expressed in terms of T.

Derivation and solution of governing equations in category (a) should be learned in your fluid mechanics course. Those in category (c) should be covered in your mass transfer course if you study chemical engineering. Combustion is a subject that generally belongs to advanced topics offered in graduate schools.

In a sense, we may state that thermodynamics is only a part of heat transfer. Nonetheless, it may be the most important part.

3-2 Solution Procedure

(a) Let us assume that the flame is subdivided by a numerical grid of 10,000 nodes. At each node, there are 3 (u, v, w) + 1 (T) + 5 (5 mass fractions) + 2 (p and ρ) = 11 variables. Hence, there are exactly 110,000 nonlinear equations.

(b) They can be, in principle, solved simultaneously by using the Newton-Raphson method. The software program nowadays is probably written in FORTRAN or C language.

(c) Among 110,000 unknowns, there are a few that are of primary interest to you, regarding how many candles you should purchase so that they will burn for two hours (for some reason you like to have candlelight throughout your party). These unknowns are the uprising velocities of melted wax inside the wick. You multiply them with wax density and the cross-sectional area of the wick to obtain the burning rate of the candle.

4. Why Is the Subject of Heat Transfer Important?

4-1. Examples of Heat Transfer Problems in Daily Life

Examples related to heat transfer in daily life abound. Let us try to follow you around in a typical morning in early December, and point out some heat transfer phenomena you may observe or encounter.

(a) Although many college students are addicted to staying up late (and therefore getting up late, and sometimes missing classes and exams), you are different. You are a very self-disciplined student, getting up regularly at 7:00 every morning. As soon as you get out of bed, you feel a little chilly. Your body loses energy by conduction from your feet at 37C to the cold floor at 20C, by convection of air flow surrounding your body, and by radiation from your skin to the wall and the ceiling. As soon as you open the curtain in your room, you will additionally lose some radiation from your skin to the outer space (assuming that the sky is cloudless). Of course, at the same time, you will also receive radiation from the wall and the ceiling, too.

(b) At 8 am, you walk to your car, which is parked on the driveway of your house near New York. You may see a layer of ice forming on the car windshield. Overnight the temperature of the glass dropped to -5C.

(c) Driving down the road, you also see an airplane flying over you. Why are your car and the airplane capable of moving? Because there is considerable amount of heat transfer going on inside the combustors. At high temperatures, air is pressurized and pushes the piston, or exiting the exhaust nozzle of the airplane, creating a thrust.

(d) On the way to the classroom, you drop by a coffee shop to buy a cup of coffee. When your hand is holding the coffee cup, it feels warm. There is heat conduction transferring from the hot coffee to your hand.

There is no need for us to go on further. There is a temperature difference, there is heat transfer flowing from hot bodies to cold bodies.

4-2 Examples of Heat Transfer Problems in Industrial Applications

(a) How did humans raise fires 4,000 years ago?

(b) Manufacturing bronze marked a significant step for humans to enter civilization. How did ancient people do that?

(c) The steam "engine" was first invented 2,000 years ago, and later improved and commercialized by James Watt in 1763.

(d) The use of the steam engine induced the industrial revolution.

(e) We wonder how George Washington spent his summers in Maryland without air conditioning.

(f) No heat transfer, no laptops that we are using. They require electronic cooling.

(g) No heat transfer, no electricity that we are enjoying. Turbine blades need to be cooled, too.

(h) No heat transfer, no cars, no airplanes, etc.

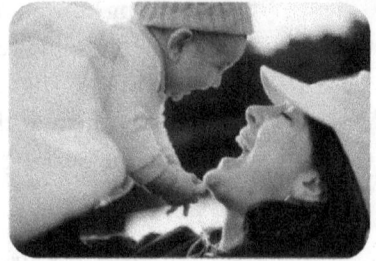

This device was invented in the first century A. D.

5. Three Modes of Heat Transfer

We mentioned conduction, convection, and radiation in the previous section. They are three distinctive modes in heat transfer.

When our hands touch an ice block, we feel cold because some energy from our hands transfers to the block. A child feels warm and loved when hugged by his mother. This phenomenon is known as conduction. Usually, heat conduction is associated with atoms or molecules oscillating in the solid. These particles do not move away from their fixed lattice positions.

When solids are replaced with fluids, such as air or water, then the phenomena become heat convection. Examples include air circulating in a room, oil flowing in a pipe, and water flowing over a plate. Hence, we may say that conduction is actually a special case of convection. When air, oil, and water stop moving, the convection problems degenerate to heat conduction.

Radiation is a phenomenon of electromagnetic wave propagation. It does not require any media to be present. A notable example is sunlight, or moonlight, reaching our earth through vast outer space, which is a vacuum. If we talk about romantic stories, the moonlight can be appropriately mentioned. In terms of heat transfer, however, we are more interested in the sunlight.

6. Prerequisites

To learn heat transfer well with this textbook, we need to know: (1) the first law of thermodynamics for both closed systems and open systems, (2) Taylor's Series Expansion, and (3) some Matlab basics. These three subjects are briefly described in the appendices of this lesson and next two.

7. Structure of the Textbook

This textbook is divided into seven major parts. Each part consists of three lessons.

Part 1 describes the introduction of the subject, three laws that govern three heat transfer modes, and one of the simplest 1-D heat conduction steady state systems without any heat generation.

In Part 2, non-zero heat generation, 1-D fins, and 2-D steady state heat conduction are analyzed. They are slightly more advanced topics than those in Part 1.

We then consider transient heat conduction problems exclusively in Part 3. Zero-D, 1-D, and 2-D problems are presented, respectively, in Lessons 7, 8, and 9.

Part 4 enters the subject of heat convection. This part should constitute the most important subject in heat transfer, and can be considered the peak of the heat transfer course during the semester. In this part, external flows are studied.

In Part 5, we focus our attention on internal flows. Some internal flow problems yield exact solutions. As in Part 4, the analyses start with hydrodynamics, and then enter the thermal aspect.

Part 6 describes application-oriented topics, including free convection, turbulent flows, and heat exchangers. They are also slightly more advanced than those in previous parts. Instructors may consider skip this part in their classes if time or students' understanding level does not permit.

Finally, Part 7 introduces basic knowledge of thermal radiation. In addition, some problems combining radiation with conduction and convection are also considered.

Such organization is also made with assigning students' exams in mind. According to the author's teaching experience at UMCP, straightforward simple grading systems may be better than complicated grading systems. The latter is exemplified by an array of activities, such as two midterms (30%), homework sets (10%), team projects (10%), quizzes (15%), presentations (5%), final exam (25%), and attendance (5%), among others.

Instead, one part is associated with one exam. Seven exams cover a semester of 14 weeks, with one exam taken every two weeks. The attendance rates tend to be high; students' burden is lowered when exams are more frequent and are of smaller scope.

Furthermore, if too much is emphasized on group work, students may have the opportunities to hitch a ride with their group mates. Or they end up sharing the work, and understanding only a small fraction of the entire group work that they are responsible for. Hence the grades become less accurate measures of their true performances in the class.

See Problem 1-3.

8. Summary

In this lesson, we have attempted to explain

(a) what the subject of heat transfer is,
(b) how it is related to thermodynamics,
(c) why it is important.

The content of the entire textbook is also outlined. Why the textbook is organized into 7 parts is justified.

9. References

1. Wikipedia.
2. Claus Borqnakke and Richard E. Sonntag, *Fundamentals of Classical Thermodynamics*, Wiley, 7th edition, 2008

10. Exercise Problems

10-1 An electrical stove in the kitchen is initially at 25C. Assume that it generates Q_g = 1.8 kW. At the same time, it loses energy to the air in the kitchen by a formula given as

$$Q\_out = 3*(T\text{-}T_\infty) \text{ in Watts.}$$

Other relevant data include: m = 1 kg, c_v = 1200 J/kg-K, and T_∞ = 25C. Take Δt = 20 sec. For simplicity, use the explicit method during time integration. Find T (t) of the stove, and plot T(t) vs. time for a period of 20 minutes.

10-2 A heat transfer system dissipates a rate that can be approximated by

$$Q = c_1 \, T^2, \text{ (in W)}$$

where c1 = 0.1 W/K^2. The system is initially at 200C. Other relevant data include: m=2 kg, c_v = 1000 J/kg-K. Take Δt = 10 sec, and use the explicit method. What is the temperature of the system at t = 10 minutes? Compare with the exact solution, too.

10-3 Which of the following states is true?

(a) The primary unknown in the equation governing conservation of energy is heat flux, q″.
(b) During the isothermal expansion process of a Carnot cycle for a piston-cylinder system, the work is found by using the first law of thermodynamics.
(c) The primary unknown in the equation governing conservation of energy is T.
(d) In an isobaric process for a piston-cylinder system containing air, $Q = c_v \Delta T$.

11. Appendix

Matlab Basics used in this textbook are limited to only simple operations. Frequently used ones are described below.

(A-1) Sum up 1, 2, 3, ..., 100

```
clc; clear
sum=0;
for i=1:100
sum = sum + i;
end
total = sum % = 5050
```

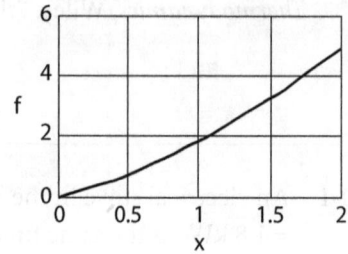

(A-2) Plot $f(x) = x^2 + \sin(x)$ versus x between 0 and 2.

```
clc; clear
L=2; nx=20; dx=L/nx; nxp=nx+1;
for i=1:nxp
x(i)=(i-1)*dx;
f(i)=x(i)*x(i)+sin(x(i));
end
plot(x, f); xlabel('x'); ylabel('f'); grid on
```

(A-3) Solve three linear equations simultaneously.

$3x + y - z = 3$
$x + 4y + 2z = 7$
$-x - 3y + 4z = 0$

```
clc; clear
a(1,1)=3; a(1,2)=1; a(1,3)=-1; b(1)=3;
a(2,1)=1; a(2,2)=4; a(2,3)=2; b(2)=7;
a(3,1)=-1; a(3,2)=-3; a(3,3)=4; b(3)=0;
q=a\b'; q' % = 1 1 1 The superscripted prime means "transpose"
```

Lesson 2
Introduction (Three Laws)

In the previous lesson, we have become aware of the definition and the importance of the subject of heat transfer. In particular, we mentioned that, in general, most heat transfer problems can be viewed as tasks of finding T(t, x, y, z). In this lesson, we will start relating q with T for all three modes in heat transfer.

Nomenclature

h = heat transfer coefficient, W/m^2-K

Q = heat transfer, J

q = heat flow rate (or heat transfer rate), W

q'' = heat flux, W/m^2

Tb = previously iterated temperature, $\bar{T}$, in K or C

α = absorptivity, between 0 and 1, if ε = α the surface is blackbody or gray

ε = emissivity, between 0 and 1

σ = Stefan-Boltzmann constant, 5.67e-8 W/m^2-K^4

1. Fourier's Law

Each of the three modes is associated with a law that relates q'' with T. We will start with conduction. In this mode, Fourier proposed, in x direction and according to the second law of thermodynamics, that

$$\text{heat flux} \propto (T_{hot} - T_{cold}). \qquad \text{(in } W/m^2) \qquad (1a)$$

We need to realize that as soon as we have accepted the validity of Expression (1a), we have used the second law of thermodynamics, depicted in Fig. 2-1. Expression (1a) is not very useful unless we modify it into an equation. Fourier [1] understood that heat

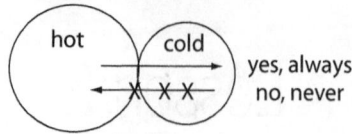

Fig. 2-1 The energy travels from a hot body to a cold body.

flux should be also inversely proportional to the distance Δx, between the two hot and cold locations. If a proportionality, k, known as thermal conductivity (with the unit of W/m-K), is further introduced, then expression (1a) becomes

heat flux = $k\,\Delta T/\Delta x$.   (1b)

According to Fig. 2-2, we should have

heat flux = k (-50)/0.1 = -500k.

Such a result is not too desirable, because the convention of our education since elementary school is that the quantity moving in positive x direction is positive. Following this convention, Fourier further inserted a minus sign and the cross-sectional area, and let Δx shrink to obtain

$$q_x = -Ak\frac{\partial T}{\partial x}\text{, (in W, called heat flow rate)} \qquad (2a)$$

or

$$q_x'' = -k\frac{\partial T}{\partial x}\text{, (in W/}m^2\text{, called heat flux)} \qquad (2b)$$

which is called Fourier's law (or Fourier's equation). It is worth noting that Eqs. (2a and 2b) can be applied to any x position in the system, and that T is not necessarily linear in x. But if Δx is taken sufficiently small, the distribution T(x) can be safely assumed to be linear within Δx.

Example 2-1

Consider a 1-D heat conduction system whose temperature has been somehow found as shown in Fig. 2-2. Relevant data are A = 0.3 m^2 and k = 2 W/m-K. What is the amount

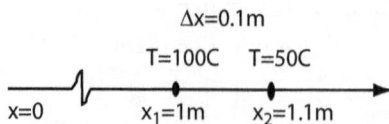

Fig. 2-2 The eastbound q" is assumed to be positive conventionally.

of heat transfer, Q in Joules, flowing from x1 to x2 during a period of one minute, if the heat flow rate is not a function of time?

Sol: Let us write the computation in a Matlab code below.

```
clc; clear
k=2; dx=0.1; A=0.3; dt=60; % properties
T1=100; T2=50; % nodal temperatures
Q=-A*k*(T2-T1)*dt/dx % = 18000 J
```

2. Law of Convective Heat Transfer

Consider fluid flows blown over both sides of a slab, as shown in Fig. 2-3. Following the sign convention that quantities moving toward the positive x direction are positive, we should assume

At x = 0, $q''_x = h_1 (T_{\infty 1} - T_1)$, (3a)

and

at x = L, $q''_x = h_2 (T_2 - T_{\infty 2})$, (3b)

regardless of the real-life situations. The unit of h is W/m^2-K.

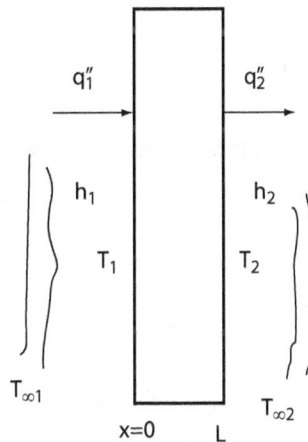

Fig. 2-3 Different versions of convective law should be taken at x = 0 and x = L.

In general, the following trends are true:

h_liquid (~200) > h_gas (~20),
h_forced_convection > h_free_convection,
h_boiling/condensation > h_single_phase.

For example, consider jumping into a swimming pool versus standing in the cold air. Our bodies are Ts = 37C and both the water temperature and air temperature are T_∞ = 5C. But we feel much colder in the water, because

h_water > h_air.

Within the subject of heat conduction, when the boundary condition is prescribed in terms of h, the value of h will usually be given. Within the subject of heat convection, however, we will learn how to find h values ourselves in Lesson 12. A few more remarks can be made:

(a) Unlike Eqs. (2a, 2b) for conduction, Eqs. (3a, 3b) can only be applied to surfaces.
(b) In y direction, we can modify Eqs. (3a, 3b) accordingly.
(c) In some literature, Eqs. (3a, 3b) are also called Newton's law of cooling.

3. Wind-Chill Factor (WCF)

When the weather is cold, we may often hear people mentioning "wind-chill factor." The term is so popular that it is beneficial for us to know a little more about it.

First, when the atmosphere is calm, let us rewrite Eq. (3b) as

$$q''_c = h_c\ (37\text{-}T_\infty), \tag{4a}$$

where the subscript "c" denotes "calm." Then, during windy conditions, let us rewrite Eq. (3b) as

$$q''_w = h_w\ (37\text{-}T_\infty), \tag{4b}$$

where the subscript "w" denotes "windy." Since $h_w > h_c$, clearly $q''_w > q''_c$. We will feel colder in the windy condition even though true air temperatures of the two cases are the same. Let us now seek a fictitious cold ambient temperature, called "wind-chill factor," such that our bodies feel q''_w with the value of h_c. Finally, the equation governing WCF becomes

$$h_c\,(37 - WCF) = h_w\,(37 - T_\infty).\qquad\qquad (5)$$

Example 2-2

Given:

$$h_c = 100 \ W/m^2\text{-}K, \ h_w = 130 \ W/m^2\text{-}K, \ T_\infty = -2C$$

Find: the value of WCF in C

Sol:

clc; clear
h_calm = 100; h_windy = 130; T_skin = 37; T_inf = -2;
q_real = h_windy*(T_skin - T_inf) % = 5070
% q_real = h_calm*(T_skin – wcf)
wcf = T_skin - q_real/h_calm % wcf = -13.7C

In the next section, we will continue to learn the law used in the radiation mode.

4. Stefan-Boltzmann Law of Radiative Emission

In radiation, the law governing the relationship between the heat flux and the temperature, called Stefan-Boltzmann law, is given as

$$q_s'' = \varepsilon\sigma T_s^4 \ ,\qquad\qquad (6a)$$

$$\text{or} \quad q_s = A\varepsilon\sigma T_s^4 \ ,\qquad\qquad (6b)$$

where ε is the emissivity ranging between 0 and 1 for all the materials on earth. Sigma, σ, is known to be the Stefan-Boltzmann constant, equal to 5.67e-8 $W/m^2\text{-}K^4$. Usually, ε is a function of the wavelength and the temperature of the surface. Before studying radiation in Lesson 19, we will assume ε to be a constant.

Example 2-3

Given: a gray surface (defined as surfaces whose radiative properties are not functions of the wavelength), Ts = 300K, and ε = 0.4
Find: the emissive radiative flux from the surface

Sol:

```
clc; clear
sig = 5.67e-8; Ts = 300; e = 0.4; A=0.3;
qflux = e*sig*Ts^4 % = 183.71 W/m^2
q = qflux*A % = 55.112 W
```

Does it imply that our bodies are constantly losing 55W to the surroundings? No, of course not. At any moment, fortunately we are also receiving radiation of similar amount emitted from the surroundings. Had we been in outer space without wearing any space suits, then the answer would be "yes".

See Problem 2-1.

5. Sheet Energy Balance

We have so far learned three laws that establish the relationships between heat flux and temperature. They are laws that are entirely unrelated to the first law of thermodynamics. In this section, we will officially apply the first law to a surface in conjunction with the three laws. This surface can be a very thin plate, or the top of a block. In either case, it is idealized as a sheet, which is too thin to possess the capability of storing any energy or receiving work. Therefore, in the first law,

$$\Delta U = Q + W, \tag{7}$$

both ΔU and W will vanish. This leaves Q to be the only surviving term. Upon closely examining the term, Q, we realize that it actually represents the **net** energy flowing into the sheet. Therefore, Eq. (7) is reduced to

$$0 = Q_{in} - Q_{out} + 0 \text{ or } Q_{in} = Q_{out}, \tag{8}$$

regardless of its being in steady state or not. An analogy to a sheet is a doorway. At t = 0, five people are about to walk through the doorway into the room. At t = 2 seconds, these five people must have walked into the room, because the doorway is too thin to accommodate people standing there. Equation (8) looks innocently trivial, but it is very important, and can confuse us sometimes.

Sometimes when we park our cars on the driveway overnight under a clear sky, in the next morning we may find that there may be a layer of ice deposited on the car windows, even though the ambient temperature of the air overnight might have stayed above the freezing point. Let us attempt to explain this interesting phenomenon below.

6-1 Clear Sky Overnight

According to Fig. 2-4, the sheet energy balance states

q_supply = q_rad + q_conv.

If we are allowed to assume that (a) there is no heater inside the car, (b) radiative exchanges between the windshield and the interior of the car are about the same, and (c) the air is fairly stagnant inside the car, such that h ≈ 0 and q_supply ≈ 0, then we have

$$\varepsilon \sigma T^4 + h(T - T_\infty) = 0, \tag{9}$$

which is a non-trivial nonlinear equation in T. Furthermore, in setting up Eq. (9), we have assumed that the windshield is hotter than the ambient air temperature, and the radiation from the outer space ($T \sim 4K$) into the windshield is negligible.

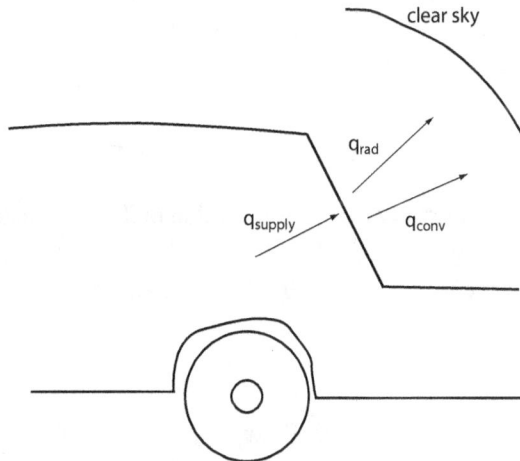

Fig. 2-4 Energy balance over the car wind shield which can be idealized as a thin sheet. A thin sheet cannot store any energy.

Example 2-4

Given: $\varepsilon = 0.4$, $T_\infty = 278K$, and h = 15.0664 W/m^2-K
Find: T_windshield

```
clc; clear
eps=.4; sig=5.67e-8; Tinf=278; h=15.0664;
% linearize T^4 = -3*Tb^4 + 4*Tb^3*T.
% better than the guess-and-check method.
% use the Newton-Raphson method.
Tb=Tinf; % initial guess
for iter=1:10
Jac=eps*sig*4*Tb^3+h;
T=(eps*sig*3*Tb^4+h*Tinf)/Jac;
Tb=T; % important to update Tb (the previously iterated temperature)
end
T % = 270 K
```

Since Ts is found to be 270K, the moisture in the air will condensate and freeze on the windshield.

See Problem 2-2.

6-2 Cloudy Sky Overnight

What if the car window is facing the cloudy sky, the wall of a building, the leaves of a tree, etc.? Then there is also radiation leaving the surfaces of those objects, arriving at the car windows. Equation (9) should be modified into

$$\varepsilon \sigma T^4 + h(T - T_\infty) - \alpha \sigma T_{surr}^4 = 0. \tag{9a}$$

Including this term will greatly increase the value of T. This inclusion also explains why sometimes car windows of one side are covered with ice layers, but those on the other side are not, because the former are facing the clear sky, and the latter are facing warm objects.

---

7. Rule of Assume, Draw, and Write (ADW)

Confusion may arise when we cannot be sure if the energy is flowing into the sheet or out from the sheet. In the case of ice formation problem, we actually had little idea if Ts was higher than T_∞ or lower prior to computations. So should we draw an arrow into the sheet, or out from the sheet? Here let us introduce a three-step rule of thumb that can help us to avoid confusion whenever deriving the sheet energy balance equation.

Step 1: Assume
Completely disregard the real-life, real-world situation. Pretend that we do not know any information even if we do. Proceed to assume a heat transfer direction. For example, let us assume that the windshield is hotter than the ambient air.

Step 2: Draw
Then, under our assumption, the arrow of q_conv should point out from the sheet, as shown in Fig. 2-4.

Step 3: Write
Finally, when writing the equation of sheet energy balance, we base on the figure drawn by us. The convection term written in Eq. (9) was written in accordance to the figure drawn.

If we adhere to this three-step ADW rule, everything will be consistent, we will not get confused, and our answers will be correct. For example, from the Matlab code, we have found that the correct answer is Ts = 270. Then our q_conv should be

q_conv = h*(Ts – T_∞) = -8h < 0,

of which the negative sign suggests that the true direction of heat convection is the opposite of our drawing.

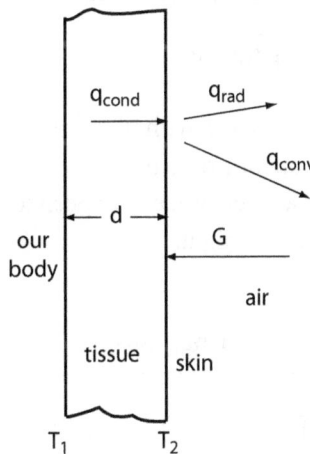

Fig. 2-5   Energy balance over the skin of our bodies

Example 2-5

In reference to Fig. 2-5, we are interested in finding the thermal conductivity of the tissue underneath our skin. Relevant data are given in the beginning of the Matlab code. Assume that T is linear in x inside the tissue.

Find: k_tissue

Let us take sheet energy balance over the skin. Thus,

$$k\_tissue * (T1 - T2)/ d = q\_rad + q\_conv - G. \tag{10}$$

Given all the necessary data, except the value of k_tissue, we should be able to find it, using Eq. (10).

Sol:

```
clc; clear
sig=5.67e-8; h=5;
e1=0.98; % see Ref[2]
d=0.0015;
T1=37+273; T2=36.5+273; Tinf=298;
G=sig*Tinf^4; % incoming radiation from surroundings
qconv=h*(T2-Tinf);
qrad = e1*sig*T2^4;
qout=qconv+qrad-G; % sheet energy balance
k=qout*d/(T1-T2) % = 0.3606
```

8. Summary

This lesson presents the following theme topics:

(a) introduce three laws that relate T with q" for three heat transfer modes,
(b) examine the wind-chill factor problem
(c) explain why sometimes a layer of ice is deposited on our car windshield if our cars are parked outdoors overnight.

9. References

1. H. S. Carslaw and J. C. Jaeger, *Conduction of Heat in Solids*, Oxford University Press, 2000.

2. http://www.thesnellgroup.com/storage/fck/File/thinkthermally/ThinkThermally_ 2007_Winter.pdf

10. Exercise Problems

2-1. In terms of radiation loss emitted, which object loses more, our bare skin or the clothes worn by us?

2-2. Given the data: $\varepsilon = 0.458$, T_inf = 278K, h= 20 W/m2-k. Find Ts by rerunning the Matlab code solving the problem of ice layer forming on the car windshield.

2-3. Revisit Example 2-5. Now, instead, the value of k is given to be 0.3606 W/m-K. But T2 is not known. Consulting the linearization technique introduced in the appendix, proceed to use the Newton method to find T2.

11. Appendix

A function $f(x+ dx)$ an be expressed in terms of $f(x)$, $f'(x)$ and $f''(x)$ in Taylor's Series Expansion as

$$f(x + dx) = f(x) + f'(x)dx + f''(x)\,(\Delta x)^2/2 + \dots \text{ higher order terms}$$

Since our interest now lies in linearization, let us retain only two terms on the right-hand side of the equation. Consequently, we have

$$f(x + dx) \approx f(x) + f'(x)dx. \tag{A-1}$$

If we imagine that T = x + dx and x = $\bar{T}$, then (T - $\bar{T}$) = dx. Equation (A-1) can be modified into

$$f(T) \approx f(\bar{T}) + f'(\bar{T})(T - \bar{T}) \tag{A-2}$$

The value of $\bar{T}$ is known from the previous iteration. At the second iteration, it is the initial condition.

If $f(T) = T^4$, we can further change Eq. (A-2) into

$$T^4 \approx -3\,\bar{T}^4 + 4\,\bar{T}^3 T,$$

which has been successfully linearized since $\bar{T}$ is just a number.

Another example is given here: $\sin(x) \approx \sin(\bar{x}) - \bar{x}\cos(\bar{x}) + \cos(\bar{x})x$.

All nonlinear terms in all nonlinear equations can be linearized in a similar fashion. Afterwards, we can write the set of equations in a matrix form as

$$[A]\{q\} = \{b\}. \qquad\qquad (A\text{-}3)$$

At the first iteration, all barred values are guessed. At the second iteration, all barred values $\bar{q}$ are updated by currently computed q via Eq. (A-3). The procedure is repeated until the solution has converged. It is known as the Newton-Raphson method.

Lesson 3
One-Dimensional Steady State Heat Conduction

When learning new things, we usually like to start from the simplest case first. Conduction is the simplest mode among the three. One-dimensional (1-D) problems are the simplest among multi-dimensions, obviously. Steady state problems are also simpler than unsteady ones.

Nomenclature

c_v = heat capacitance, or specific heat, J/kg-K
d_0, d_1, d_2, d_3 = undetermined constants, K or K/m
h1 = heat transfer coefficient on the left face of a slab, W/m^2-K
k = thermal conductivity, W/m-K
L_t = total thickness of a multi-layer slab, La + Lb, m
qL = heat flux at the left face of the slab
qR = heat flux at the left face of the slab
R = overall thermal resistance, (W/m^2-K)$^{-1}$

1. Governing Equation for T(x) or T(i)

Within this slab, let us take a thin slice as our control volume, and index its center as i, shown in Fig. 3-1. If the problem satisfies the three conditions, i.e., (a) 1-D, (b) steady state, and (c) no heat generation, then the heat flux entering the slice at *w* should be equal to that exiting at *e*. That is,

$$q_w - q_e = 0.$$

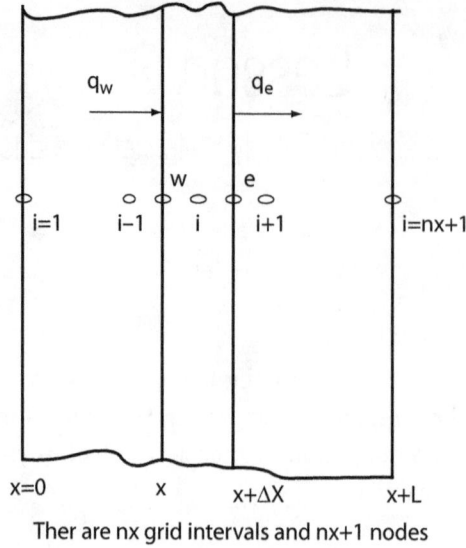

Ther are nx grid intervals and nx+1 nodes

Fig. 3-1   One-D Slab Discretized with a Grid

If they are not equal, the inequality implies that there will be some decrease or increase of energy in the slice. Then heat conduction in the slice does not qualify the status of steady state. Hence,

$$k(T(i\text{-}1) - T(i))/\Delta x = k(T(i) - T(i+1)) /\Delta x,$$

or $T(i\text{-}1) - T(i) - T(i) + T(i+1) = 0,$

or $T(i) = 0.5*(T(i\text{-}1) + T(i+1)).$ (1)

If the temperature distribution satisfies Eq. (1), it must be a linear function in x. The proof of this linearity can be made if we start with three nodes first. It is obvious that the line connecting two arbitrary values of $T(i\text{-}1)$ and $T(i+1)$ will pass $T(i)$, if $T(i)$ is the average of them. Successively, we can extend the same argument to higher numbers of nodes.

Armed with the knowledge that $T(x)$ is linear in x, we are able to proceed to solve for $T(x)$ below.

See a Matlab code solving Eq. (1) by solving a set of linear equations from i=2 to i=10.

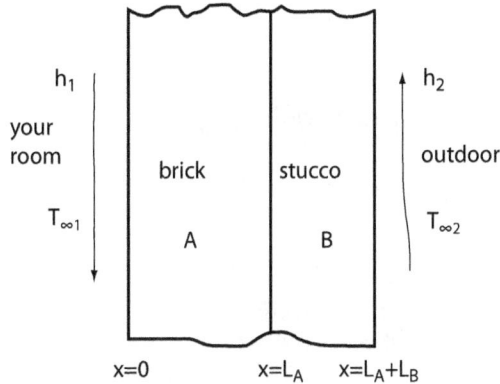

Fig. 3-2   System schematic of 1-D two-layer slabs

2. A Single-Slab System

Consider the wall of your house separating the outdoor ambient airflow and the indoor room-temperature airflow, as shown in Fig. 3-2. Let us first consider the case without the stucco layer. Figure 3-2 can still be referenced.

Since T(x) is linear in x, we can confidently write

$$T(x) = d_0 + d_1 x, \tag{2}$$

with two undetermined constants. Next, we take the sheet energy balance at x=0 to yield

$$h_1 \left(T_{\infty 1} - T(0) \right) = -k \left(\frac{dT}{dx} \right)_{x=0},$$

which can be subsequently manipulated to become

$$h_1 d_0 - k d_1 = h_1 T_{\infty 1}. \tag{3a}$$

Similarly, we can derive a sheet energy balance equation at x=La as

$$h_2 d_0 + (h_2 L + k) d_1 = h_2 T_{\infty 2}. \tag{3b}$$

Let us now resort to a Matlab code to solve these two linear equations for d_0 and d_1 below. The room temperature and the outdoor air temperature are 25C and 2C, respectively.

```
clc; clear
h1=20; h2=40; k=1.4; L=0.2;
```

```
Tinf1=25; Tinf2=2;
a(1,1)=h1; a(1,2)=-k; b(1)=h1*Tinf1;
a(2,1)=h2; a(2,2)=h2*L+k; b(2)=h2*Tinf2;
d = a\b;
T1=d(1) % = 19.7213C
T2=d(1)+d(2)*L % = 4.6393C
% check the global energy balance
qL = h1*(25-T1) % = 105.574 W/m^2
qR = h2*(T2-2) % = 105.574 W/m^2
```

Rarely are we interested in knowing T1 and T2. However, as soon as we know their values, we can quickly find heat fluxes. The heat flux flowing from the warm room to the slab should be the same as that flowing from the slab to the outdoor atmosphere.

See Problems 3-1, 3-2, and 3-3.

3. A Two-Slab System

When there are two layers of different materials, it is convenient for us to write two linear profiles, as

$$T_A(x) = d_0 + d_1 x, \tag{4a}$$

$$T_B(x) = d_2 + d_3 x, \tag{4b}$$

At x=La, both the temperatures and the heat fluxes are continuous, yielding two more conditions. Together with two conditions at x=0 and x=La + Lb, we have four conditions for four unknowns, d_0, d_1, d_2, and d_3.

```
clc; clear
h1=20; h2=40; ka=1.4; La=0.2; kb=0.7; Lb=0.06; Lt=La+Lb;
a(1,1)=h1; a(1,2)=-ka; b(1)=25*h1;
a(2,1)=1; a(2,2)=La; a(2,3)=-1; a(2,4)=-La; b(2)=0;
a(3,2)=-ka; a(3,4)=kb; b(3)=0;
a(4,3)=h2; a(4,4)=h2*Lt+kb; b(4)=2*h2;
d = a\b;
T1=d(1) % = 21.2118C
T2=d(1)+d(2)*La % = 10.3882C
T3=d(3)+d(4)*Lt % = 3.8941C
% check the global energy balance
```

$qL = h1*(25-T1)$ % $= 75.7647$ W/m^2
$qR = h2*(T3-2)$ % $= 75.7647$ W/m^2

thermal ~ electrical

See Problem 3-4.

Making analogies is nice.

4. Three or More Slabs

In principle, we can continue to extend straight-forwardly the analysis shown in Sec. 3. If we potentially have 10 slabs, there will be 20 unknowns. Correspondingly, we have two boundary conditions at x=0 and x=L_t, together with 18 continuity conditions of T and q'' at 9 interfaces.

If our interest is just limited to finding the heat flux, then we can consider using the circuit analogy alternatively. In terms of Fig. 3-2, let us write:

$$q'' = h_1 (T_{\infty 1} - T(0)) = k_a [T(0) - T(L_a)]/L_a = k_b [T(L_a) - T(L_t)]/L_b = h_2 (T(L_t) - T_{\infty 2})$$

Breaking them apart, and aligning all the terms yields

$q''/h1 = T_{\infty 1} - T(0)$

$q''La/ka = T(0) - T(La)$

$q''Lb/kb = T(La) - T(Lt)$

$q''/h2 = T(Lt) - T_{\infty 2}.$

Adding them up, we obtain

$$q'' R = T_{\infty 1} - T_{\infty 2}, \tag{5}$$

$$\text{where R} = \frac{1}{h_1} + \frac{L_a}{k_a} + \frac{L_b}{k_b} + \frac{1}{h_2}. \tag{6}$$

Equations (5, 6) take the analogous form of $I*R = V$ in the electrical circuit. Once q'' is calculated, T(0) can be found from

$$T(0) = T_{\infty 1} - q''/h1,$$

and T(La) can be found from

$$T(La) = T(0) - q''L_a/k_a$$

successively.

Example 3-1

Use the circuit analogy to find qL in Section 3, and check if the values are the same.

```
% use the circuit analogy
R =1/h1 + La/ka + Lb/kb + 1/h2 % = 0.3036
qL = (25-2)/R % = 75.7647 W/m^2, same as that in Sec. 3.
```

5. Severe Restrictions Imposed by Using Electrical Circuit Analogy

For Eq. (5) to be valid, there are at least three conditions we should meet: (1) the system must be 1-D; (2) the system must be in steady state; and (3) there should not any heat generation inside any of the slabs. These restrictions are quite severe. Therefore, let us not overemphasize this topic.

6. Other Types of Boundary Conditions

In addition to the convective boundary condition are two simpler ones. They are

(a) At x = 0, T = 100C. This is known as the boundary condition of Dirichlet type, which prescribes the function value.
(b) At x = 0, dT/dx = -0.5. This is known as the boundary condition of Neumann type, which prescribes the derivative of a function. It is equivalent to the specification of heat fluxes.

It is worth noting that a 1-D steady-state problem cannot be subject to Neumann boundary conditions on both sides. We will leave this topic as an exercise problem.
See Problem 3-5.

In total, we can possibly have 8 (9 – 1) pairs of boundary conditions, which are tabulated below:

(a) T1 + T2, (b) T1 + q2, (c) T1 + convective BC at 2,

(d) q1 + T2, (e) q1 + convective BC at 2,
(f) convective BC at 1 + T2, (g) convective BC at 1 + q2, (h) convective BC both sides.

7. Thermal Properties of Common Materials—Table 1

	k (W/m-K)	ρ (kg/m^3)	c_v (J/kg-K)	ν (m^2/sec)
Aluminum	237	2,702	903	n/a
Wood	0.17	540	1,385	n/a
Oil	0.15	884	1,909	5.50e-4
Water	0.61	1,000	4,180	8.55e-7
Air	0.026	1.16	717.5	1.60e-5
Liquid Metal	8.54	13,529	139.3	1.13e-7

These properties can be frequently referred to when we work on problems in later lessons.

8. Summary

This lesson presents analyses of 1-D steady-state heat conduction in a single slab and multi-slabs. The classical circuit analogy is also mentioned, but not emphasized. Readers are encouraged to run Matlab codes in Secs. 2 and 3 to enhance their understanding of 1-D heat conduction.

In particular, readers are also advised to work on the exercise problems.

9. References

1. Wikipedia: thermal conductivity of air (or water)
2. P. J. Schneider, *Conduction Heat Transfer*, Addison-Wesley Publishing Co., 1955.

3-1 Revisit the single-slab problem in Sec. 2. Set h1 to be zero. Without running the Matlab code or deriving the analytical result, what do you predict values of T1 and T2 to become, by your physical intuition?

3-2 Revisit the single-slab problem in Sec. 2. We have been taught that q" is proportional to k of the slab. Do you expect to see a linear profile of q" versus k, for $10 < k < 800$?

3-3 Revisit the single-slab problem in Sec. 2. Find Ts analytically if the slab reduces to a sheet.

3-4 Revisit the single-slab problem in Sec. 2. Nondimensionalize the boundary condition on the right face (x=L) by introducing $x^* = x/L$ and $\theta = (T - T_{\infty 1})/(T_{\infty 2} - T_{\infty 1})$. You should obtain a very important parameter.

3-5 If the heat fluxes on both sides are equal to, for example, 100 W/m2, with k=0.1 and L=0.1m, then T1=101C and T2=1C can be the solution; T1=750C and T2=650C can be the solution, too. The solutions will be infinitely many.

On the other hand, if , the problem by nature cannot remain steady state.

11. Appendix

Given: T1= 0C, T11 = 100C, L = 0.1m, and nine internal nodes, steady state, shown in Fig. 3-3.

Find: T2, T3, T4,..., T10

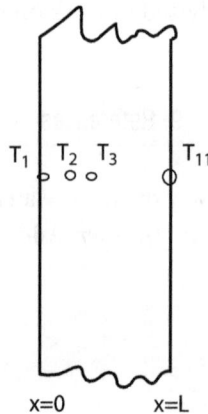

$$T_1 \quad T_2 \quad T_3 \qquad T_{11}$$

x=0   x=L

Fig. 3-3   A grid system of 11 nodes

Sol:

This exercise is a very standard basic heat conduction problem, which can be conveniently solved using a Matlab code.

```
%%
clc; clear
nx=10; nxp=nx+1; L=0.1; dx=L/nx; % in this problem, dx is not used.
T(1)=0; T(nxp)=100;
% It is more convenient to intentionally include the boundary condition.
a=zeros(nxp);
a(1,1)=1; b(1)=T(1);
for i=2:nx
a(i,i-1)=1; a(i,i)=-2; a(i,i+1)=1; b(i)=0;
end
a(nxp,nxp)=1; b(nxp)=T(nxp);
T = a\b'; T'
0 10.0000 20.0000 30.0000 40.0000 50.0000 60.0000 70.0000 80.0000 90.0000
100.0000
```

as expected.

Lesson 4

One-Dimensional Slabs with Heat Generation

In this lesson, let us study problems of 1-D slabs that are embedded with heat generation. Related issues will also be examined and discussed.

Nomenclature

A = cross-sectional area, m^2

bi = Biot number, defined as h $\Delta x / k$

Q_g^* = a convenient term for heat generation, defined as, $q_{gen}'''(\Delta x)^2 / k$, K

s1 = $(kb\, \Delta xa)/(ka\, \Delta xb)$

---

1. Introduction

Heat generation sources are commonly exemplified by electrical resistance, nuclear reaction, and chemical reaction, of which a special case is the metabolism in our bodies. Thermal energy is generated as the final output that tends to increase the temperature of the system. Inversely, there are also endothermic chemical reactions that will behave as heat depletion sinks.

The main reasons that we cannot treat heat generation and heat fluxes in the same fashion are at least twofold:

(a) Heat generation takes place **inside** the control volume, whereas heat fluxes occur **across** the boundary of the control volume.

(b) Partly due to reason (a), heat generation in a 2-D system is usually a function of t, x, y, and T(i, j), but does not involve neighboring nodal temperatures.

By contrast, heat fluxes are expressed by T(i, j) as well as neighboring nodal temperatures.

Finally, analyses of systems containing heat generation cannot be conducted using the electrical analogy, because the heat flux is now a function of x, and is no longer a constant, as in zero heat generation cases. The existence of heat generation makes the global energy balance more interesting, too.

2. Governing Equations

Let us start with the first law of thermodynamics applied over a slice of control volume, represented by Δx, shown in Fig. 4-1. The law states

$$\Delta U = Q + W. \tag{1}$$

Since the system is in steady state, we have $\Delta U = 0$. The heat flow rate into the system can be written as

$$Q = (q_w - q_e)\Delta t,$$

where $q_w = kA(T_{i-1} - T_i)/\Delta x$, and $q_e = kA(T_i - T_{i+1})/\Delta x$.

Let us treat the heat generation as a form of work. Thus,

$$W = q_{gen}''' A\Delta x \, \Delta t.$$

It is observed that there is A in all three terms. So it can be conveniently dropped. This convenience is partly why we focus on the Cartesian coordinates. Had the cylindrical coordinates been used, A would have been a function of r.

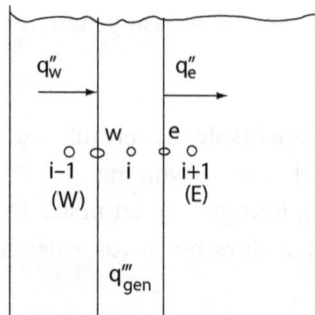

Fig. 4-1   The control volume for a 1-d slab with heat generation

Substituting all terms into Eq. (1) and dropping leads to

$$-T_{i-1} + 2T_i - T_{i+1} = Q_g^* \qquad (2)$$

which is the governing equation for T(i) in the present problem, with the term, Q_g^*, defined in the nomenclature. Equation (2) can be applied to interior nodes only. Governing equations for temperature at boundary nodes should be derived using heat conservation over control volumes of a half grid interval width, or simply using the sheet energy balance.

See Problem 4-1.

3. Heat Conduction Related to Our Bodies

We all care about our own bodies. And we should. Without health, nothing matters. Let us study heat conduction related to them first.

3-1 Estimate the Heat Generation of Our Bodies

Because of metabolism, our bodies constantly generate thermal energy, which is lost to the surroundings at the same rate, keeping our body temperature at constant 37C. Sometimes at night during sleep, we may feel too hot. That is because our bodies generate too much energy, which is trapped by the blanket that covers our bodies.

It will be interesting for us to estimate the heat generation of our bodies. Taking the sheet energy balance at the surface of clothes worn by us schematically shown in Fig. 4-2, we can write

$$k_{air}(37 - T_{fabric})/L_{air} = h(T_{fabric} - T_\infty). \qquad (3a)$$

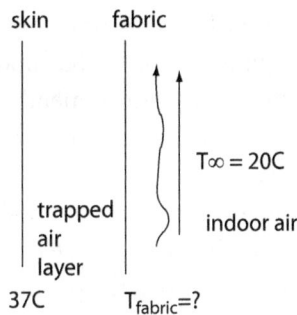

Fig. 4-2   The system schematic of our skin covered by a layer of fabric

Solving for T_{fabric}, we obtain

$$T_{fabric} = (37 + BiT_\infty)/(1 + Bi),$$ (3b)

where Bi is the Biot number, defined as $Bi = hL_{air}/k$. It is worth noting that usually k in Bi is the thermal conductivity of the solid. Here the trapped air is treated like a solid. The following Matlab code first computes the temperature of fabric, idealizing our bodies as cylinders.

Example 4-1

Estimate the heat generation of our bodies. Relevant data are given in the code.

Sol:
```
clc; clear
Tinf=20;
m=70; k_air=0.026; L_air=0.003; h=5;
Bi=h*L_air/k_air;
Tf=(37+Bi*Tinf)/(1+Bi) % = 30.78C ~ 31C
%
R_body = 0.12;
H_body = 1.76;
V= H_body*(pi*R_body^2/4);
A_body = H_body*(2*pi*R_body);
Q_conv = A_body*h*(Tf - Tinf) % = 71.53 W
q_gen = Q_conv/V % = 3.6e3 W/m^3
```

According to our computation, we find that volumetric heat generation rate of our bodies is approximately 3.6 kW/m^3. In reality, the correct value may be even higher if we take into account our perspiration [1]. So, our bodies are equivalently leaving a 100-Watt light bulb on all the time. In any case, remember not to lick your body to cool down yourself during a dance party, as the kangaroo tried to cool down itself in the photo shown.

It may be an unpleasant thought. But when the body heat generation ceases, the body temperature will decrease as time elapses. Detectives usually can measure the temperature of a corpse to estimate how long the person has been dead.

See Problem 4-2.

Finally, the radiation emitted by our clothed bodies and the radiation received by them approximately cancel each other out. Our bodies without clothes on, however, will lose more radiation than they receive, according to:

$$\sigma(310K)^4 \approx 524W \text{ emitted versus } \sigma(298K)^4 \approx 447W \text{ received,}$$

assuming an emissivity being nearly unity.

3-2 A Coarse Grid System

We will now turn our attention to finding $T(x)$ in a heat-generating system. You are going out camping in a cabin that is wired for electricity but is not equipped with central heating. Overnight the outdoor temperature, especially in the wilderness, can be quite low. You are debating if you should bring a plain blanket or an electric blanket with you.

Let us use a coarse grid system. It is also a good idea to purposely include the boundary condition T1=37C in the computation for the sake of convenient programming.

As shown in Fig. 4-3, there are six grid nodes. The set of six equations can be written in the matrix form as:

T1	T2	T3	T4	T5	T6	b'	
[1	0	0	0	0	0;	= 37C	
-1	2	-1	0	0	0;	= 0	
0	-1	1+s1	-s1	0	0;	= 0	
0	0	-1	2	-1	0;	$= Q_g^*$	
0	0	0	-1	2	-1;	$= Q_g^*$	
0	0	0	0	-1	1+bi;];	$= bi^* T_\infty$	

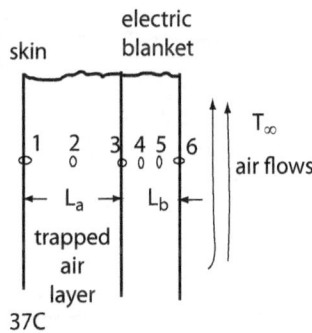

Fig. 4-3   A system of 1-D two-layer slab with heat generation inside the electrical blanket

or b = [37 0 0 Q_g^* Q_g^* bi*T_∞], which is the transpose of b' shown above. The Matlab code can be written to compute the linear system using T=a\b'.

```
clc; clear
Tinf=-5; h=10;
ka=0.026; La=0.001; dxa=La/2;
kb=0.04; Lb=0.002; dxb=Lb/3;
s1=kb*dxa/(ka*dxb); % convenient quantity
bi=h*dxb/kb; % Biot number
qgenb =2e5;
Qgs =qgenb*dxb*dxb/kb; % convenient quantity
a=[1      0      0      0    0    0;
  -1      2     -1      0    0    0;
   0     -1    1+s1    -s1   0    0;
   0      0     -1      2   -1    0;
   0      0      0     -1    2   -1;
   0      0      0      0   -1  1+bi;];

b = [37  0  0  Qgs  Qgs  bi*Tinf];
T = a\b';
T(3) % = 28.43C (qgen off) 35.23C (qgen on)
T(6) % = 17.26C (qgen off) 26.27C (qgen on)
% There is a difference of approximately 7 degrees between using the plain blanket and
using an electric blanket.
```

See Problems 4-3 and 4-4.

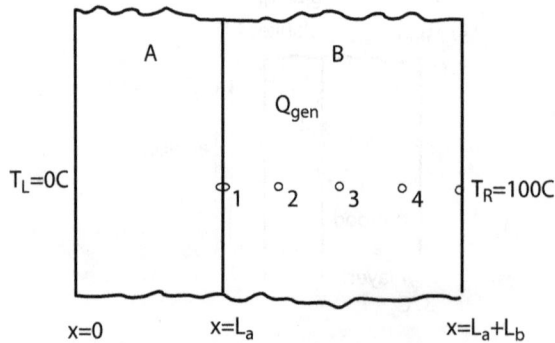

Fig. 4-4   A two layer system for the optimization problem

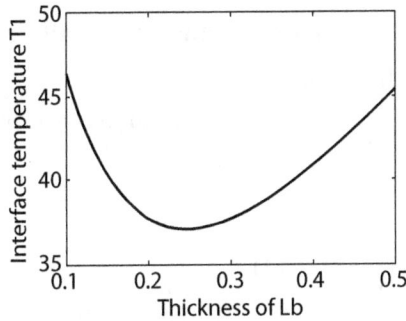

Fig. 4-5   The interfacial temperature versus the thickness of L$_b$

3-3 Optimization

Optimization is an interesting problem. Living our lives is a typical optimization process. We would all like to be able to enjoy our lives by partying all night and vacationing in Hawaii all year long. Unfortunately, we also need to study or/and work to support ourselves and our families.

Now, consider a 1-D slab of materials A and B, as shown in Fig. 4-4. At $x = 0$, $T_L = 0C$, is given. At $x = L_a$, is not known, but we desire to minimize it by covering slab A with slab B to prevent T1 from becoming too high. Intuitively, the thicker Lb is, the better insulation there is. However, if there exists a heat source embedded inside slab B (like an electric blanket), then the thicker Lb is, and the more energy will be generated by the heat source. Therefore, the problem becomes unclear to us unless we run numerical experiments.

Example 4-2

Given: TL = 0C, TR = 100C, ka = 10 W/m-K, kb = 7.3 W/m-K, La = 0.1m; vary L_b between $0.1m \leq L_b \leq 0.5m$, $q'''_{gen,b}$ = 0.2e5 W/m^3. Take four grid intervals inside slab B, as shown in Fig. 4-4.

Find: the optimal such that is at its minimum.

Sol: See the Matlab code in the appendix. We will present only the computed result in Fig. 4-5. Indeed, we observe that there exists a minimum for T1. If we choose L_b to be about 0.24 m, the minimum T_1 value is obtained. Needless to say, when k and $q'''_{gen,b}$ vary, the curves will vary, too.

4. Discussions

In addition to topics mentioned above, a few issues are worth our special attention.

4-1 Radiation Boundary Condition

If we account for the radiative emission at $x = x_6$ in Fig. 4-3, the equation governing T6 should be modified into

$$k_b(T_5 - T_6)/\Delta x_b = h(T_6 - T_\infty) + \epsilon\sigma T_6^4,$$
$$(4)$$

which is a nonlinear equation. It can be linearized using Taylor's Series Expansion, and solved together with other linear equations iteratively.

Would you like to play with me?

I would, but... I am too tired right now...

This radiative activity should not directly affect equations governing other nodal temperatures. In fact, in the case where T6 is prescribed, whether or not there is radiation taking place at x = x6 should not affect the solution of the problem at all. We need to keep this in mind.

4-2 Sketch the Trends and Check our Speculations with Running the Matlab Code

It is a good habit to take advantage of the availability of the computer code by running numerical experiments. What parameters can we vary in the body heat generation problem? There are La, Lb, ka, kb, $q'''_{gen,b}$, and h. Some physical trends are obvious. For example, if we decrease T_∞, all nodal temperatures should decrease, assuming other parameters remain the same. Some may not be too obvious as demonstrated in Example 4-2. A few general rules of thumb can be suggested:

(a) Vary only one parameter at a time. Do not try to vary two parameters or more at the same time. We will get confused.

(b) Make sure the energy is balanced globally. This checking procedure offers a soothing assurance that our code is likely bug-free.

(c) It is perfectly fine to ask others or ourselves wild or nonsense questions. We are not born to ask good questions. Gradually, however, questions we raise will become increasingly better. The author used to be an active table tennis player. In the beginner stage, whenever he asked other players in the club, "Would you like to play with me?" the answer often was, "Ah, no, I guess not; I am too tired." We all need to go through a somewhat humiliating stage in order to attain advanced stages. If we feel bashful in asking bad questions, we will be forever unable to ask good questions.

See Problem 4-5.

4-3 Check the Global Energy Balance

In the body heat generation problem in Sec. 3-2, for example, let us compute energy flowing into the system, which is the sum of two components: heat conduction at x= 0 and the heat generation in the blanket. Then the energy out from the system takes place at x = La + Lb.

Qin = ka*(37-T(2))/dxa + qgenb*2*dxb % = 312.653 W
Qout = h*(T(6)-Tinf) % 312.653 W

Note that, in slab B, we only take two interior nodes, 4 and 5. The presence of only two nodes is the reason we multiply qgenb in the code by 2, even though slab B contains three grid intervals. As we increase nx to 1000, for example, missing one grid interval should not be a concern for us.

5. Summary

We have studied systems containing heat generation inside 1-D slabs. Essential concepts include:

(a) Do not use the electrical circuit analogy any more.
(b) Running Matlab codes can help us to understand physical trends.
(c) An optimization problem related to minimizing the interfacial temperature is analyzed. The critical thickness of the insulation layer was determined. For Lb > 0.24m, there is too much heat generation. For Lb < 0.24m, the insulation is too thin to provide insulation effects.
(d) A few issues were discussed.

6. Reference

1. http://en.wikipedia.org/wiki/Thermoregulation

7. Exercise Problems

4-1 Based on Eq. (2), prove that T(x) is quadratic in x if the heat generation is constant.

4-2 The problem related to the wisdom of detectives who are able to estimate how long a corpse has been dead is beyond the scope of our learning at this moment. Comment on why.

4-3 In the problem of the electric blanket, derive the governing equation for T(3) if, instead of taking sheet energy balance, we now take a control volume of half grid interval, $0.5\Delta x$.

4-4 What is the value of qgen in the blanket, so that the air trapped inside the blanket can function as a perfect insulator? In other words, what is the value of qgen when T(3)=37C?

4-5 If the heat generation is turned off, and if T1 and T2 on both faces of a *steady-state* 1-D slab are given, then T(x) and q"(x):

    (a) do not vary when k varies,

    (b) vary when k varies,

    (c) may vary if the right face is blown over by a hot airflow,

    (d) should vary if the right face is blown over by a hot airflow, and if suddenly h varies.

8. Appendix

```
clc; clear % optimize the value of Lb to achieve the goal of minimal interface temperature
La = 0.1;
ka = 10; kb = 7.3; Qg = 0.2e5;
TL = 0; TR = 100;
Lb1=0.1; Lb2=0.5; dLb = (Lb2-Lb1)/20; % vary Lb from 0.1m to 0.5m
for iLb = 1:21
Lb = Lb1+(iLb-1)*dLb;
Lplot(iLb)=Lb; % stored for plotting only
dx = Lb/4; c1 = (kb/ka)*(La/dx);
c2 = Qg*dx*dx/kb;
a(4,4) = 0; b(4) = 0;
a(1,1) = 1+c1; a(1,2) = -c1; b(1) = TL;
a(2,1) = 1; a(2,2) = -2; a(2,3) = 1; b(2) = -c2;
a(3,2) = 1; a(3,3) = -2; a(3,4) = 1; b(3) = -c2;
a(4,3) = 1; a(4,4) = -2; b(4) = -c2-TR;
T = a\b'; Tplot(iLb) = T(1);
end
plot(Lplot, Tplot);
xlabel('thickness of Lb'); ylabel('interface temperature T1')
hold off
```

% *T1m = min(Tplot) % = 37.125C*
% *Index = find(Tplot <=T1m) % at iLb = 8*
% *Lb = Lb1 + (Index-1)*dLb % Lb = 0.24*

Lesson 5

One-Dimensional Steady-State Fins

Fins are extended surfaces for the purpose of heat transfer enhancement, mostly cooling. One-D steady-state fins represent those in which the temperature is only a function of x, but not a function of r (or y) or t. These systems are different from those one-D slabs in that there exists an extra term due to heat convection in the governing equation for interior nodes.

Nomenclature

$c_1 = hp(\Delta x)^2/(kA_c)$

$c_2 = \sigma \varepsilon p(\Delta x)^2/(kA_c)$

p = perimeter of the fin, m ($p = \pi D$ for circular fins)

T_b = temperature of the fin at the base, C

T_m = average temperature of the fin, defined as, $(\int_0^L T dx)/L$, K or C

β = a dimensionless parameter similar to the Biot number, $hpL^2/(kA_c)$

ξ = dimensionless x, defined as x/L

Fins are installed on Radioisotope Thermoelectric Generator (RTG) for cooling effects by radiation in outer space. Photo taken on the moon.

1-1 Main Purpose of Fins

Planar or circular fins are objects extended from, generally, a hot base as shown in Fig. 5-1. The main purpose of a fin is to cool a hot surface by increasing the contact surface area between the hot base and the cold fluid.

Unlike shark fins that are designed for better swimming performance, fins in the subject of heat transfer are designed to increase the cooling rate. A very basic understanding of fin performance is that the cooling rate is supposed to increase after the installment of the fin.

Without the presence of the planar fin, the surface area of the base (to be attached by a fin) is merely $A_c = Z*d$. Thus, the cooling rate will be

$$q_1 = Zd\, h_1\, (T_b - T_{\infty 1}). \tag{1a}$$

After the fin is installed, the cooling rate will become

$$q_2 = 2ZLh_2\, (T_m - T_{\infty 2}), \tag{1b}$$

where the mean temperature of the fin, Tm, is lower than Tb, h2 will be lower than h1 due to extra extrusion of the fin, and $T_{\infty 2}$ may be higher than $T_{\infty 1}$ due to crowdedness if several fins are installed. Therefore, the basic understanding is that Equation

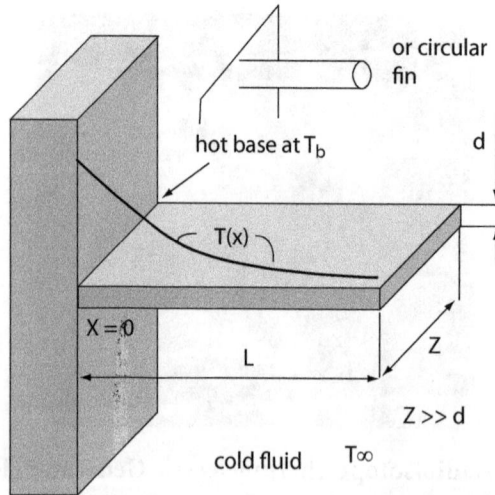

Fig. 5-1   Typical fins for cooling of the hot base

(1b) should yield higher q than Eq. (1a) should. And we cannot infinitely increase the number of fins and expect to increase q continually, either.

See an optimization problem in the Appendix.

Example 5-1

Find a case where q1 = q2 if T_b and $T_{\infty 1}$ are 100C and 20C, respectively. Therefore, there is little thermal-aspect achievement for us to install a fin.

Sol:

Set $2Lh_2 (T_m - T_{\infty 2}) = dh_1 (T_b - T_{\infty 1})$. If:

(a) L/d = 20 based on the heat-exchanger design,
(b) h2/h1 = 0.5 due to obstruction of fins to the airflows,
(c) Tm = 30C due to poor conductivity of the fin, and
(d) T_∞ increases from 20C to 26C due to crowdedness of fins,

then the equation above is satisfied. We have wasted money and time, but accomplished nothing.

1-2 Difference Between 1-D Slabs and 1-D Fins

The main difference lies in the governing equations. In 1-D slabs, the governing equation contains only two fluxes, one at the left face and the other at the right face. If there is heat generation, it exists throughout the entire control volume. In 1-D fins, however, in addition to the two heat fluxes, there is also convective flux crossing the circumferential area. The contrast can be seen clearly from the sketch in Fig. 5-2. When $h \to 0$, the fin problem does reduce to a 1-D slab problem.

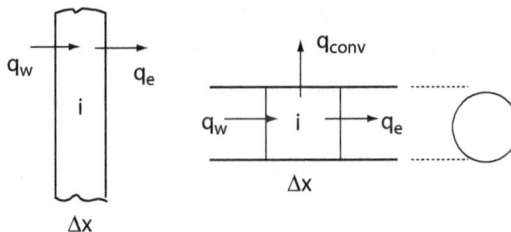

Fig. 5-2   Energy balance over a slice of the fin

1-3 A Quick Estimate

If there is no drastic change near the base of the fin—namely, k is relatively large and h is relatively small—perhaps we can use a quick and approximate way to solve the problem. Since this method seems to cultivate students' analytical skills, let us present it below.

Let us assume T(x) to be a quadratic profile in x as

$$T(x) = d_0 + d_1\xi + d_2\xi^2, \tag{2a}$$

where $\xi = x / L$.

At x = 0, T(0) is equal to Tb. We can quickly modify the profile into

$$T(x) = T_b + d_1\xi + d_2\xi^2, \tag{2b}$$

There are two conditions we can use to determine d1 and d2.

At x = L, or $\xi = 1$, let us assume that the tip of the fin is insulated. Hence, we have

$$d_1 + 2d_2 = 0. \tag{3a}$$

The second condition is very important. It is dictated by the global energy balance. In other words, the energy flowing out from the base at x=0 to the fin should be equal to the convective loss from the entire fin to the surrounding air. In the dimensional form, we write

$$-kA_c\left(\frac{dT}{dx}\right)_{x=0} = hp\int_0^L (T - T_\infty)\,dx,$$

which can be straightforwardly transformed into

$$\left(1 + \frac{\beta}{2}\right)d_1 + \left(\frac{\beta}{3}\right)d_2 = -\beta(T_b - T_\infty), \tag{3b}$$

where β is a dimensionless parameter, defined in the nomenclature. Equations (3a, 3b) can be solved simultaneously to yield the temperature profile. The accuracy of this method is low, as expected. Nonetheless the result allows us to observe some qualitative trends.

For higher accuracies, it is better for us to resort to higher-order polynomials or numerical solutions. In the appendix, a cubic-profile temperature distribution is assumed and compared with the numerical solution.

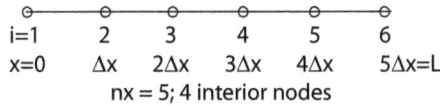

Fig. 5-3　A coarse 1-D grid system

---

2. Analyses

2-1 Governing Equation

In reference to Fig. 5-2, we have, for a steady-state fin,

$$q_w = q_e + q_{conv},\tag{4}$$

where

$$q_w = kA_c\,(T_w - T(i))/\Delta x,\ q_e = kA_c\,(T(i) - T_E)/\Delta x,\ \text{and } q_{conv} = hp\Delta x\,(T_i - T_\infty)$$

The final result, after some algebraic manipulation, becomes

$$-T(i - 1) + (2 + c_1)T(i) - T(i + 1) = c_1 T_\infty,\tag{5}$$

for i = 2, 3, 4, …, nx interior nodes. The Matlab for loop can be used to generate these equations conveniently:

```
for i = 2:nx
a(i,i-1)=-1; a(i,i)= (2+c1); a(i,i+1)=-1; b(i)= c1*T_inf;
end
```

For example, for nx=5, there are 5 grid intervals, 4 interior points, and 6 total grid points, as shown in Fig. 5-3.

2-2 A Standard Matlab Code Solving T(x) and for a Fin

A standard Matlab code is presented below, solving a plain fin problem with data given in the code. The solution T(x) is shown in Fig. 5-4.

```
clc; clear
k=400; L = 0.5; D=.004; p=pi*D; Ac=D*D*pi/4; % properties of fin
h=100; T_inf=20; % condition of the ambient air
```

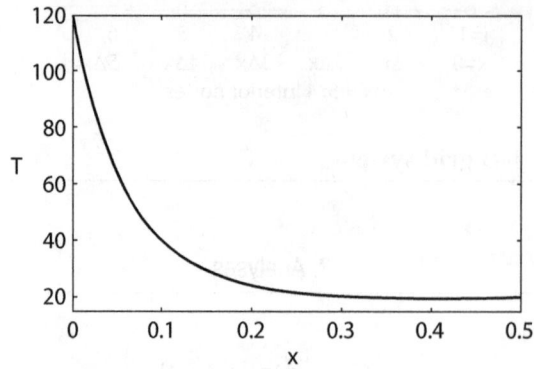

Fig. 5-4 The computed temperature distribution of a 1-D fin

---

nx=150; dx=L/nx; dxs=dx*dx; nxp=nx+1; % grid system
c1=h*p*dxs/(k*Ac); % convenient or important parameters
Tb=120; % boundary condition at the base
a=zeros(nxp); b=zeros(1,nxp); % establish zero elements for the sparse matrix.
x=linspace(0,L, nxp);
% establish x coordinates
a(1,1)=1; b(1)=Tb; % node 1
for i=2:nx
a(i,i-1)=1; a(i,i)=-(2+c1); a(i,i+1)=1; b(i)=-c1*T_inf; % interior nodes
end
a(nxp,nx)=-1; a(nxp,nxp)=1; % The tip of the fin is insulated.
T=a\b';
%
plot(x,T); axis([0 L 15 120]);
xlabel('x'); ylabel('T'); hold off
qb = k*Ac*(T(1)-T(2))/dx % = 7.7410;

%
%qb3 = k*Ac*(1.5*T(1)-2*T(2)+ 0.5*T(3))/dx;
% Note that a three-point approximation can yield better accuracy, if so desired

2-3 Equivalent Governing Differential Equation

Equation (5) can be algebraically rearranged into

$$T(i-1) - 2T(i) + T(i+1) - c_1\,(T(i) - T_\infty) = 0$$

Keeping in mind that

$[T(i-1) - 2T(i) + T(i+1)]/(\Delta x)^2 \approx \dfrac{\partial^2 T}{\partial x^2}$ we can obtain

$$\frac{\partial^2 T}{\partial x^2} - d_1(T - T_\infty) = 0, \qquad (6)$$

where $d_1 = hp/(kA_c)$ [1].
See Problem 5-2.

---

3. Fins Losing Radiation to Clear Sky Overnight

Equation (6) can be analytically solved to yield exact solution of exponential functions. However, if there is a slight deviation from the ideal case, such as variable k or Ac, or inclusion of radiation loss, then it may be very difficult for the analytical method to solve the problem. By contrast, for numerical methods, sometimes a small modification of the formulation, thus the code, is all we need to enable us to solve problems that cannot otherwise be solved analytically.

A notable example is a fin cooled by both convection and radiation. Equation (5) can be modified to include radiation to become

$$-T(i-1) + (2 + c_1)T(i) - T(i+1) = c_1 T_\infty - c_2 [T(i)]^4,$$

where c_2 is defined in the nomenclature. We can place the radiation term on the right-hand side of the equation to try our luck. If the solution has converged, we are lucky. The job is done. If the solution has diverged, we can linearize the nonlinear term using the Newton-Raphson method.

% Fins Exposed to Clear Sky Overnight (Radiation Loss)
% Clear sky at night emits very little radiation
clc; clear
k=400; D=.004; h=150; e=0.8; sig=5.67e-8;
T_inf=20+273; Tb=120+273;
Ac=D*D*pi/4; p=pi*D;
Lx=0.5; nx=150; dx=Lx/nx; dxs=dx*dx;
nxp=nx+1;
c1=h*p*dxs/(k*Ac); c2=e*sig*p*dxs/(k*Ac);
%
a=zeros(nxp); b=zeros(1,nxp); % establish the base for the sparse matrix.
x=linspace(0,Lx, nxp);

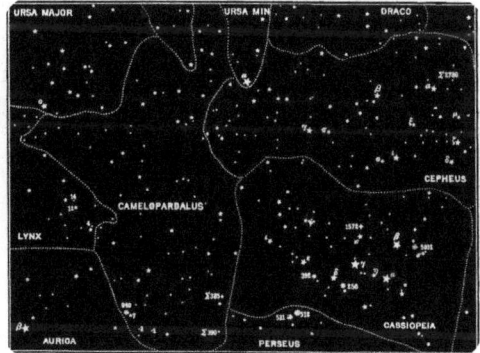

```
for i=2:nx; T(i)=Tb; end % initial guess
for iter=1:7
a(1,1)=1; b(1)=Tb;
for i=2:nx
a(i,i-1)=1; a(i,i)=-(2+c1); a(i,i+1)=1; b(i)=-c1*T_inf+c2*(T(i))^4;
end
a(nxp,nx)=-1; a(nxp,nxp)=1; % The tip of the fin is insulated.
T=a\b;
% monitor the convergence trend
fprintf('%9.4f %9.4f %9.4f \n', T(2), T(nx/2), T(nxp))
end
plot(x,T); axis([0 Lx 15+273 120+273]); hold off
xlabel('x'); ylabel('T(x) in K');
text(0.1, 360,'fin exposed to clear sky overnight')
qb = k*Ac*(T(1)-T(2))/dx % = 9.8208
```

The solution has converged. To check the global energy balance involving radiation is slightly more challenging than without radiation. This task will be left as an exercise problem. See Problem 5-3.

4. A Seemingly Puzzling Phenomenon

As k increases, for example, from 300 W/m-K to 400 W/m-K, we do expect the heat flow rate at the fin base, q_b, to increase. In fact, hypothetically, if we can find on Earth a certain material, whose k approaches infinity, we should expect T(x) of the fin to become Tb uniformly throughout, because there is no thermal resistance inside the fin.

A puzzling question thus arises: if T(x) becomes a constant, which is Tb, then don't we have dT/dx = 0? With infinite k and zero temperature gradient, the cooling rate at the base,

$$q_b = -kA_c \left(dT/dx \right)_{x=0},$$

therefore will become

$$q_b = \infty * 0.$$

What value should it be? To answer this question correctly, we need to examine which quantity increases faster, k or (dT/dx)_base. Let us find out from the following table computed using the Matlab code above.

k	$(dT/dx)_{x=0}$	q_b
300	2154.2869	8.1215
310	2120.5394	8.2607
320	2088.3476	8.3977
330	2057.5951	8.5326
340	2028.1773	8.6655
350	2000.0000	8.7965
360	1972.9779	8.9255
370	1947.0340	9.0528
380	1922.0980	9.1784

It can be seen that k increases faster than $(dT/dx)_{x=0}$ decreases, thus the net product, q_b, increases as k increases.

In fact, it is possible to prove analytically that q_b is proportional to $\sqrt{k}$ under the boundary condition of insulation at x = Lx.

See Problem 5-2, again.

5. Fin Efficiency

In the industry and in the literature, the fin efficiency [2] is defined as

$$\eta_f = q_b / q_{max},$$

where q_{max} is the value that can possibly attain when k of the fin is allowed to change. When k of the fin approaches infinity, q_{max} is attained, according to our analysis above. Thus,

$$q_{max} = hp \int_0^L (T(x) - T_\infty) dx = hpL (T_b - T_\infty).$$

Example 5-2

What is the fin efficient described in the Matlab code in Sec. 2-2?

Sol: *qmax=h*p*L*(Tb-T_inf);*
efficiency = qf/qmax % = 12.32%

6. Optimization

In this section, we will examine two optimization problems below.

6-1 Constraint of Fixed Total Volume

If the total volume (LA_c) of a fin is fixed, does there exist a critical fin length such that the maximum cooling rate exists? By our intuition, as we increase the length of a fin, the surface area seems to increase (ignoring the tip surface area). See the table below:

	L (m)	D (m)	V (m^3)	$A_{circumf}$
1	0.25	0.08	0.001257	0.0628
2	1	0.04	0.001257	0.1257

or the formula:

$$A_{circumf} = pL = (\pi D) \left[\frac{V}{0.25\pi D^2} \right] = \frac{4V}{D}$$

On the other hand, as L increases, T_m decreases as the fin material is located far away from the base. Therefore it is possible that there exists a critical value of L such that q_b is maximized. Indeed, after computations, we find that q_b reaches its maximum for L ≈ 0.065 m, as shown in Fig. 5-5.

See the Matlab code in the Appendix for this optimization problem.

6-2 Optimization for Rod Bundles

Intuitively, in reference to Fig. 5-6, if we increase the number of fins, we should expect to increase the cooling heat flow rate from the hot base to the cool air. However, if n_fin is too large, such that fins are too crowded, it is possible that h may decrease, and T_inf may increase, reducing the cooling flux. Let us see if there exists an optimal number of fins such that the cooling rate is maximized.

Assume empirically that

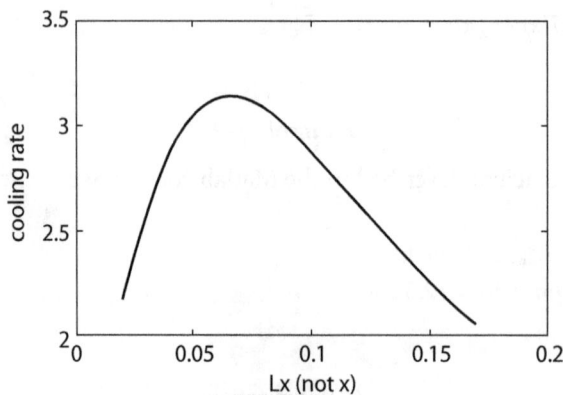

Fig. 5-5   Optimization of the thickness of a slab

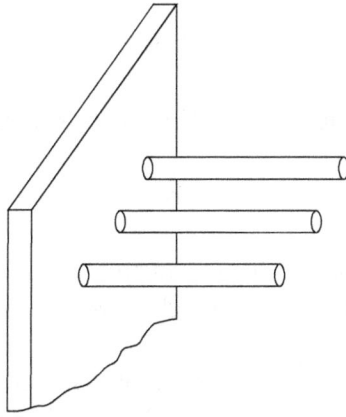

Fig. 5-6   Optimization of the number of fins

$$h = c_1 - c_2 n^2,$$

where $c_1 = 8.0188$ and $c_2 = 0.0188$, and that

$$T_\infty = 20 + (n - 1)^{2.4},$$

where n is number of fins. Relevant data are given in the code.

We have found that $n_{critical} \approx 12$, as shown in Fig. 5-7.

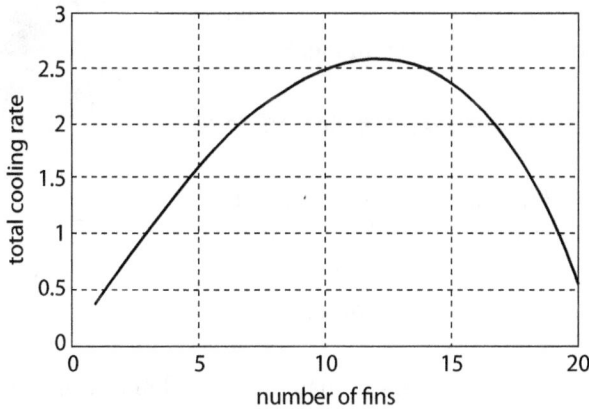

Fig. 5-7   Total cooling rate versus the number of installed fins

7. Summary

In this lesson, the subject of fins is given. Key points include:

(a) A quick estimate of the temperature profile can be made by assuming the profile to be quadratic.
(b) A standard Matlab code solving the fin problem is presented.
(c) We attempt to explain a seemingly puzzling phenomenon related to k, $(dT/dx)_{x=0}$, and q_b.
(d) The fin efficiency and some discussions of trends are mentioned.
(e) We solved an optimization problem related to finding the number of fins such that the cooling effect of a fin bundle is at its maximum.

8. Reference

1. Theodore L. Bergman, Adrienne S. Lavine, Frank P. Incropera, and David P. Dewitt, *Fundamentals in Heat and Mass Transfer*, Wiley, 7th edition, 2011.
2. D. Q. Kern and A. D. Kraus, *Extended Surface Heat Transfer*, McGraw-Hill, 1972.

9. Exercise Problems

5-1 Your mom is cooking spaghetti to reward you for having studied heat transfer hard. The lid handle of the pot is made of stainless steel whose thermal properties are given below. Let us approximate the curved handle as a straight fin, of which both ends are attached to the lid, as shown in Fig. 5-8.

k=40; L = 0.15; D=.004; p=pi*D; Ac=D*D*pi/4; % properties of fin
h=20; T_inf=20; % condition of the ambient air
nx=50; dx=L/nx; dxs=dx*dx; nxp=nx+1; % grid system
c1=h*p*dxs/(k*Ac); % convenient or important parameters
Tb=120; % boundary condition at the base

Find T(0.5L). Explain to your mom that it is safe to hold the lid handle at the center for a few seconds, even though the lid itself may be as hot as 120C.

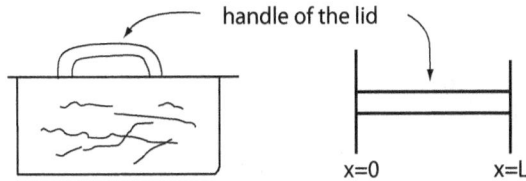

Fig. 5-8 The system schematic of the handle of the lid, approximated as straight rod

5-2 Make an attempt to show that q_b is proportional to $\sqrt{k}$ analytically. Equation (6) may be helpful.

5-3 Add a small block of statements in the Matlab code given in Section 3 to compute energy entering the fin and energy leaving the fin, including radiation, and to check if the global energy balance is valid.

5-4 Vary h and k in a standard fin problem. Do you expect T(x) profile to shift up or down? Run numerical experiments to confirm with your own predictions.

10. Appendix

A-1. Explanation of the Seemingly Puzzling Phenomenon

The Matlab code in an attempt to explain the seemingly puzzling phenomenon is given below. The result is given in the text.

```
clc; clear
for ik=1:11
k=300+(ik-1)*10;
D=.004; Ac=D*D*pi/4; p=pi*D;
h=150; T_inf=20; Tb=120;
Lx=0.5; nx=150; dx=Lx/nx; dxs=dx*dx; nxp=nx+1;
c1=h*p*dxs/(k*Ac);
a=zeros(nxp); b=zeros(1,nxp); % establish the base for the sparse matrix.
x=linspace(0,Lx, nxp);
a(1,1)=1; b(1)=Tb;
for i=2:nx
a(i,i-1)=1; a(i,i)=-(2+c1); a(i,i+1)=1; b(i)=-c1*T_inf;
end
a(nxp,nx)= -1; a(nxp,nxp)=1; % The tip of the fin is insulated.
T=a\b';
```

```
%plot(x,T); axis([0 Lx 15 120]); hold off
dTdxn = (T(1)-T(2))/dx;
qf2 = k*Ac*dTdxn;
fprintf('%7.0f %9.4f %9.4f \n', k, dTdxn, qf2)
end
```

A-2 Cubic-Polynomial Temperature Profile

The analysis for cubic polynomial T profile is given below. Let us assume

$$T(\xi) = T_b + d_1\xi + d_2\xi^2 + d_3\xi^3.$$

Three conditions are needed to determine d_1, d_2, and d_3. The first two conditions are similar to those imposed for quadratic profiles. The third condition states that the heat flow rate at x=0 should be equal to the sum of the heat flow rate at x=0.5L and heat convection loss from the fin to the cooling air between x=0 and x=0.5L. That is,

$$-kA_c\,(dT/dx)_{x=0} + kA_c\,(dT/dx)_{x=0.5L} = hp(0.5L)(T_m - T_\infty)$$

or

$$-(dT/d\xi)_{\xi=0} + (dT/d\xi)_{\xi=\xi^*} = \beta\xi^*(T_m - T_\infty),$$

where $T_m = [\int_0^{\xi^*} (T_b + d_1\xi + d_2\xi^2 + d_3\xi^3)\, d\xi]/\xi^*$

In this analysis, ξ^* is taken to be 0.5. Other values, that are less than 1, can be taken, too. The algebraic equation can be readily written by inspecting the coefficient matrix in the Matlab code.

See Fig. 5-9 for the results of two temperature distributions, and comparisons of them.

```
clc; clear
k=400; D=.004; Ac=D*D*pi/4; h=20;
% At h=2, quadratic model is still good. But not at h=20
Lx=0.4; nx=200; dx=Lx/nx; dxs=dx*dx; nxp=nx+1;
p=pi*D; c1=h*p*dxs/(k*Ac);
Tb=120; Tinf=20;
a=zeros(nxp); b=zeros(1,nxp); % establish the base for the sparse matrix.
x=linspace(0,Lx, nxp);
a(1,1)=1; b(1)=Tb;
```

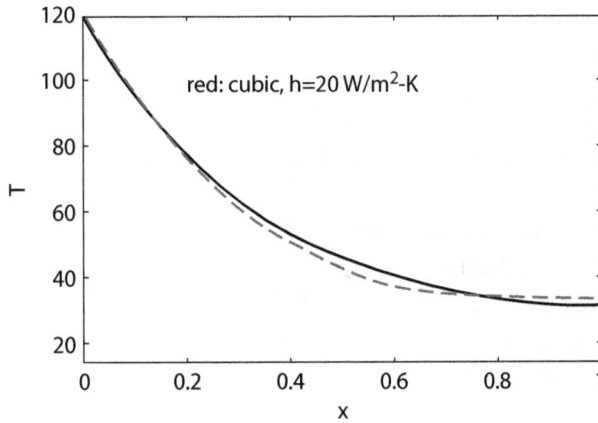

Fig. 5-9 Comparisons of two temperature distributions obtained by the cubic-profile model and the numeric solution

```
for i=2:nx
a(i,i-1)=1; a(i,i)=-(2+c1); a(i,i+1)=1; b(i)=-c1*Tinf;
end
a(nxp,nx)=-1; a(nxp,nxp)=1; % The tip of the fin is insulated.
T=a\b';
% check global energy balance
sum=0;
for i=2:nx
sum=sum+h*p*dx*(T(i)-Tinf);
end
qbase = k*Ac*(T(1)-T(2))/dx %
qconv = sum %
%
be=h*p*Lx*Lx/(k*Ac); dT=Tb-Tinf; e1=0.5;
% >>>>> quadratic polynomial
a=[1 2; 1+be/2 be/3]; b=[0 -be*dT];
d=a\b';
Tm = Tb + d(1)/2 + d(2)/3
% >>>>> cubic polynomial
aa=[1 2 3; 1+be/2 be/3 be/4;
 e1*e1*be/2 e1^3*be/3-2*e1 e1^4*be/4-3*e1^2];
bb=[0 -be*dT -e1*be*dT];
d=aa\bb';
Tm = Tb + d(1)/2 + d(2)/3 +d(3)/4
% >>>>>
```

```
for i=1:nxp
 kxi(i)=x(i)/Lx;
 % Tpoly(i)=Tb+kxi(i)*d(1)+kxi(i)^2*d(2);
 Tpoly(i)=Tb+kxi(i)*d(1)+kxi(i)^2*d(2)+kxi(i)^3*d(3);
end
plot(x/Lx,T, kxi, Tpoly, 'r--'); axis([0 1 15 120]);
xlabel('x'); ylabel('T'); hold off
text(0.2, 100, 'red: cubic, h=20 W/m^2-K')
qbase2=-k*Ac*d(1)/Lx
qconv2=h*p*Lx*(Tm-Tinf)
```

A-3 Constraint of Fixed Fin Mass (or Volume)

```
clc; clear
Tb=120; T_inf = 20; h = 50;
k=250; D=.003; Ac=D*D*pi/4; Lx =0.12; Vt = Ac*Lx;
dLx = 0.005;% increment of fin-length change
for iL = 1:31
Lx = 0.02 + (iL-1)*dLx; Lxplot(iL) = Lx;
Ac=Vt/Lx; D = sqrt(Ac*4/pi); p=pi*D;
nx=20; dx=Lx/nx; dxs=dx*dx; nxp=nx+1;
c1=h*p*dxs/(k*Ac);
%
a=zeros(nxp); b=zeros(1,nxp); % establish the base for the sparse matrix.
x=linspace(0,Lx, nxp);
a(1,1)=1; b(1)=Tb;
for i=2:nx
a(i,i-1)=1; a(i,i)=-(2+c1); a(i,i+1)=1; b(i)=-c1*T_inf;
end
a(nxp,nx)=-1; a(nxp,nxp)=1;
% The tip of the fin is insulated.
T=a\b';
%plot(x,T); axis([0 Lx 15 120]); hold off
dTdxn = (T(1)-T(2))/dx;
qf = k*Ac*dTdxn; qfplot(iL)= qf;
end
plot(Lxplot, qfplot)
xlabel('Lx (not x)'); ylabel('cooling rate')
```

A-4 Fin Bundles and Optimization

```
clc; clear
% h = c1 - c2*n^2, where c1 = 8.0188 and c2 = 0.0188
% T_inf = 20 + (n-1)*2.4,
k=30; D=.003; Ac=D*D*pi/4; Lx = .12;
nx=20; dx=Lx/nx; dxs=dx*dx; nxp=nx+1; p=pi*D;
Tb=120;
c1 = 8.0188; c2 = 0.0188;
for n=1:20
nv(n)= n;
T_inf = 20 + (n-1)*2.4; h = c1-c2*n^2;
e1=h*p*dxs/(k*Ac);
a=zeros(nxp); b=zeros(1,nxp); % establish the base for the sparse matrix.
x=linspace(0,Lx, nxp);
a(1,1)=1; b(1)=Tb;
for i=2:nx
a(i,i-1)=1; a(i,i)=-(2+e1); a(i,i+1)=1; b(i)=-e1*T_inf;
end
a(nxp,nx)=-1; a(nxp,nxp)=1; % insulated at the tip
T=a\b';
%plot(x,T); axis([0 Lx 15 120]); hold off
dTdxn = (T(1)-T(2))/dx;
qf = k*Ac*dTdxn; qfplot(n)= n*qf;
end
plot(nv, qfplot); grid on % n ~ 12
xlabel('number of fins'); ylabel('total cooling rate')
```

Lesson 6

Two-Dimensional Steady-State Conduction

After we have learned 1-D steady state heat conduction, logically the next topic can be extended to 2-D steady state heat conduction. Generally, as shown in Fig. 6-1, heights of 2-D systems are no longer as tall as those of 1-D systems, which walls of our houses resemble. The depth, Z, of a 2-D system should still be much larger than both Lx and Ly. In a 3-D system, like an elephant, Lx, Ly, and Lz are all comparable.

Nomenclature

$e = 0.5\ \Delta r$, m

q'''_{gen} = volumetric heat generation rate, W/m^3

$Q^*_g = q'''_g (\Delta x)^2/k$, K

r_1 = grid interval ratio, $(\Delta x/\Delta y)^2$

Z = depth of the control volume, m

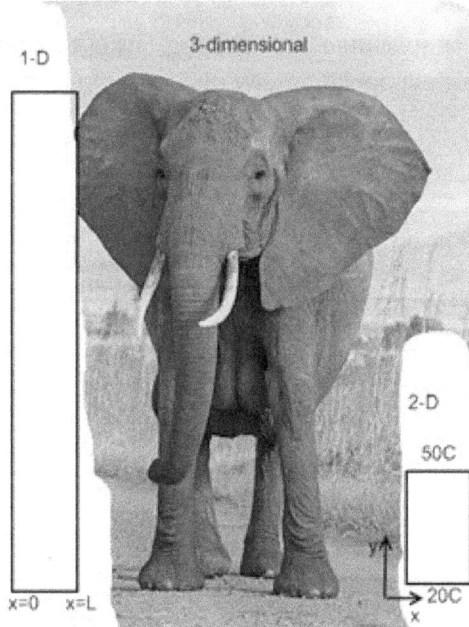

Fig. 6-1   Comparisons of 1-D, 2-D, and 3-D systems.

1-1 Derivation of the General Governing Equation

The first law of thermodynamics for a closed system states

$$\Delta U = Q + W,$$

where $\Delta U = 0$ because the system is in steady state. Also,

$$W = \Delta x\, \Delta y\, Z q_g'''\Delta t,$$

and

$$Q = (q_w - q_e + q_s - q_n)\Delta t.$$

In reference to Fig. 6-2, the heat flow rate can be expressed in terms of the nodal temperatures according to Fourier's law as

$$q_w = k\, \Delta y\, Z(T_W - T_{ij})/\Delta x, \; q_e = k\, \Delta y\, Z(T_{ij} - T_E)/\Delta x,$$

$$q_s = k\, \Delta x\, Z(T_S - T_{ij})/\Delta y, \text{ and } q_n = k\, \Delta x\, Z(T_{ij} - T_N)/\Delta y.$$

All these terms can be substituted into the first law of thermodynamics. Then after straightforward algebraic manipulations, we obtain

$$-T_W + (2 + 2r_1)\, T_{ij} - T_E = r_1\,(T_S + T_N) + Q_g^* \tag{1}$$

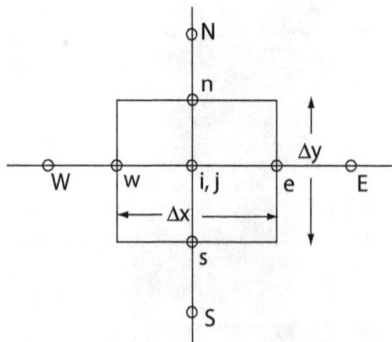

Fig. 6-2   A typical 2-D computational cell

Note that "W, E, S, and N" nodes are, respectively, "i-1, j", "i+1, j", "i, j-1", and "i, j+1" in the Matlab codes. They look friendlier than indices i and j in the text. Furthermore, we intentionally place TS and TN on the right-hand side of Eq. (1) because, with small modifications, we can continue to use codes that solve 1-D systems. The method is also known as the line-by-line solver, which is generally more stable than the point-by-point Gauss-Seidel method, and converges faster, too.

A code using the Gauss-Seidel point-by-point method is also given in the appendix.

1-2 Three-Interior-Node System

Next, let us apply Eq. (1) to a coarse grid system with only three nodes, so that we will not be immediately overwhelmed by the Matlab coding.

Example 6-1

Where should we place a point heat generation such that the mean temperature of the system, as shown in Fig. 6-3, is the maximum?

```
clc; clear
% T1 = 100C, find the location of q_gen is such that Tm is maximum
k=5; qgen=2e4;
dx=0.1; dxs=dx*dx; dy=0.1;
r1=(dx/dy)^2; r22=2+2*r1;
qs2=0*qgen*dxs/k; qs3=qgen*dxs/k; qs4=0*qgen*dxs/k;
% locate the point source at node 3
% convenient quantities
T(1)=100; T(5)=0;
for iter=1:200
a=[1 0 0 0 0;
 -1 r22 -1 0 0;
 0 -1 r22 -1 0;
 0 0 -1 r22 -1;
 0 0 0 0 1 ];
b=[T(1) 2*r1*T(2)+qs2 2*r1*T(3)+qs3 2*r1*T(4)+qs4 T(5)];
```

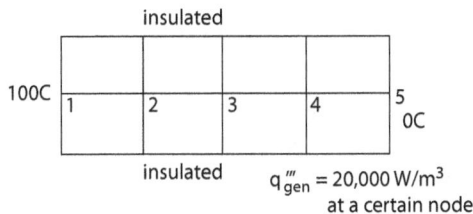

Fig. 6-3   A 2-D conduction system of 3 interior nodes

```
% We purposely left TS and TN on the right-hand side
% for the purpose of code modifications later
qq=a\b'; T=qq;
end
T'
% check the global energy balance
qin = k*dy*(T(1)-T(2))/dx + qgen*dx*dy % = 225 W
qout = k*dy*(T(4)-T(5))/dx % = 225 W
Tm = sum(T(2:4))/3 % = 76.67 C
```

It can be readily shown that, if we move the point heat generation source to node 2 or node 4, the mean temperature becomes 70C. We can speculate that the heat source should be installed as far away from the boundaries as possible, so that more nodes can benefit from the energy distributed by the heat source.

We are aware of the analogy between thermal resistance and electrical resistance. In fact, an analogy between temperature in a conduction system and population of people in an area can be recognized, too. Possibly, if a county planning commissioner likes to see higher uniform population increases in the county, he may propose to build a theme park near the center of the county.

See Problem 6-1.

So far, we have consistently prescribed one boundary condition of a certain type at a boundary node for every node. Is it possible, instead, to give more than one boundary condition at a node, while retrieving the boundary condition from another node?

Example 6-2

Over-Specification of the Boundary Conditions

In reference to Fig. 6-4, let us over-specify the boundary condition at x=0, by having T1 = 100C and q1 = 80W, but treating T5 as an unknown. Find the temperature distribution. Also note that the top face is no longer entirely insulated to avoid triviality of the problem.

The key is to establish the governing equation for T5, using the condition q1=80W. If the entire top face is insulated, we should have

$$q_1 = 80 = q_5 = k\Delta y(T_4 - T_5)/\Delta x.$$

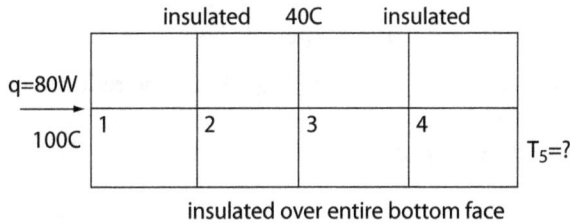

Fig. 6-4   Similar to Fig. 6-3, but over-specified boundary condition on the left
            face

In case of the system shown, we should have, instead

$$q_1 = k(T_3 - 40) + k(T_4 - T_5). \tag{2}$$

which can be viewed as the governing equation for T5, and can be readily implemented into the Matlab code.

```
% specify both q1 and T1 at node 1
clc; clear
k=5; qgen=0; qflux1=80; TN=40;
dx=0.1; dxs=dx*dx; dy=0.1; r1=(dx/dy)^2; r22=2+2*r1;
qs=qgen*dxs/k;
T(1)=100;
T=100*ones(1,5); % initial guess
for iter=1:200

a=[1 0 0 0 0;
 -1 r22 -1 0 0;
 0 -1 r22 -1 0;
 0 0 -1 r22 -1;
 0 0 1 1 -1];
b=[ T(1) 2*r1*T(2) r1*(T(3)+TN) 2*r1*T(4) TN+qflux1/k];

q=a\b'; T=q;
end
T'
% check the global energy balance
qin = qflux1 % = 80 W
qout = k*dy*(T(4)-T(5))/dx+k*dx*(T(3)-TN)/dy % = 80 W
```

1-3 A Special Case

If $\Delta x = \Delta y$, and there is no heat generation, Eq. (1) can be reduced to a simple form:

$$T(i,j) = 0.25 \, (T_W + T_S + T_E + T_N). \tag{3}$$

When we sell our house, the appraiser usually assumes the value of our house to be the average of values in the neighborhood, using a formula very similar to Eq. (3), known as the Laplace Equation when $\Delta x = \Delta y$.

Equation (3) is the governing equation for an arbitrary nodal temperature in the interior region. If there are totally n nodal unknowns, there are exactly n recursive equations, no more, no fewer. Note that, according to the finite difference formulation we take for the energy balance, *nw*, *sw*, *se*, and *ne* corner nodes are not related to the node (i,j). In some finite element analyses, however, these nodal temperatures may also enter the stencil energy balance [1].

We should further note that ρ and cv are properties associated with time-dependent problems, and should not appear in the steady-state governing equation.

2. A Standard Matlab Code Solving 2-D Steady-State Problems

In the appendix, we present a Matlab code that solves a standard 2-D steady-state heat conduction problem. We can modify it for problems subject to various boundary conditions. Again, we need to keep in mind that a specification of all Neumann boundary conditions around the boundary of a 2-D system will lead to either an unsteady solution or a set of infinitely many non-unique solutions.

Whenever possible, it is a good habit for us to check if the code yields solutions that satisfy global energy balance for the system. In the examples above we did conduct such checks.

See Problems 6-2, 6-3, and 6-4.

3. Maximum Heat Loss from a Cylinder Surrounded by Insulation Materials

Even though we have not emphasized, and are not going to emphasize, the use of cylindrical coordinates in this textbook, it is beneficial for us to be aware of the difference between the Cartesian coordinates and the cylindrical coordinates, and of the discretization formulation of the latter.

3-1. Governing Equation

In reference to Fig. 6-5, we can write

$$q_s = 2k\pi\Delta x \, [r(j) - e](T_S - T_j)/\Delta r,$$

$$q_n = 2k\pi\Delta x \, [r(j) + e](T_j - T_N)/\Delta r,$$

where $e = 0.5\Delta r$. If the system is 1-D and in steady state, we should have

$$q_s - q_n = 0,$$

or, after straightforward algebra,

$$(r(j) - e)T_S - 2r(j)T_j + (r(j) + e)T_N = 0. \tag{4}$$

We can readily show that Eq. (4) is equivalent to the differential equation

$$r\frac{d^2T}{dr^2} + \frac{dT}{dr} = 0.$$

3-2 An Optimization Problem

We can use Eq. (4) to optimize the radius of a layer of the insulation material, such that we attempt to avoid the maximum heat loss from the cylinder to the surrounding air.

Shown in Fig. 6-5, the thickness of the insulation material, $R_2 - R_1$, is the variable to be optimized. If the thickness is too small, the insulation effect is not sufficiently

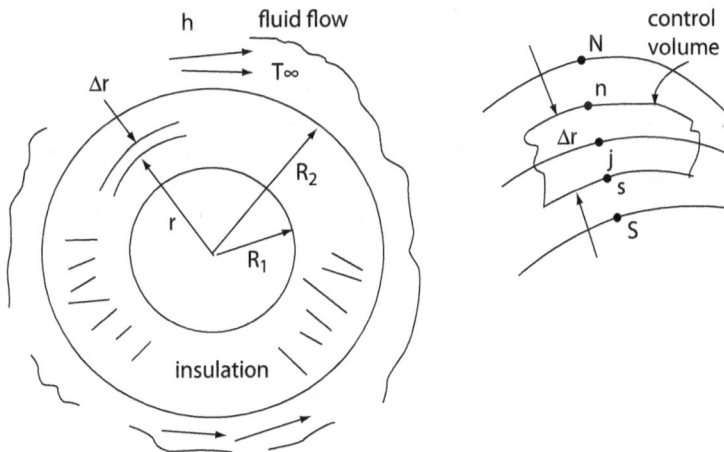

Fig. 6-5   A typical computational cell in the radial direction

pronounced. On the other hand, if the thickness is too large, the area of the material increases, tending to lose a large amount of energy via convection.

See the Matlab code in the appendix for the numerical procedure. The final result of heat loss versus the thickness of insulation is shown in Fig. 6-6 for a particular set of data including k, h, R1, and R2.

The peak seems to appear only for small values of R1. Above these values, as we increase the insulation, the heat loss monotonically decreases.

---

4. Summary

In this lesson, we have studied 2-D steady-state heat conduction problems. Key topics include:

(a) derivation of the governing equation

(b) a standard Matlab code computing 2-D steady-state heat conduction problems is presented.

(c) a three-node system, from which we can learn where to place a point heat source such that Tm of the system is maximum

(d) one-D heat conduction in the cylindrical coordinates is also presented, so that we become aware of the formulation involving variable r.

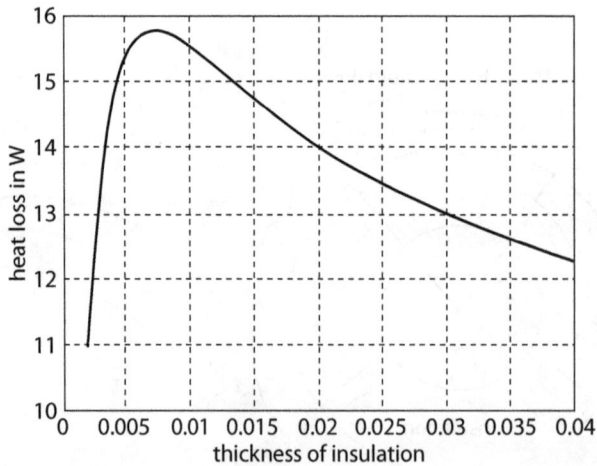

Fig. 6-6   **Determination of the insulation thickness at which the heat loss is maximum**

---

5. References

1. T. M. Shih, *Numerical Heat Transfer*, Springer-Verlag, 1984.
2. http://en.wikipedia.org/wiki/Diagonally_dominant_matrix

6. Exercise Problems

6-1 Consider a 2-D system (for a total of eleven nodes in the x direction, but still three nodes in the y direction), using the standard 2-D code. Do you draw the same conclusion stated in Example 6-1?

6-2 Which value of T_m is higher for the two cases of square systems shown in Fig 6-7? Predict first and run the numerical experiments. Find T_m to see if your prediction is correct. Take nx = ny = 8.

6-3 In reference to Fig. 6-7, change the right boundary condition (of the right figure) to a convective boundary condition with $T_\infty = 50C$, h=40 W/m^2-K, k = 2 W/m-K, nx = ny =3, and Lx = Ly = 0.3m. (So, T on the right boundary is no longer 50C!) Find six nodal temperatures. Then double the values of h and k. Observe if your solution changes. If not, comment on why not.

6-4 Which one yields more energy output from the system to the surroundings, in association with Problem 6-3?

6-5 Consider only two grid nodes in a rectangular 2-D system as shown in Fig. 6-8. Find the energy supply (in W) required to maintain the 2-D system at steady state for both cases. Take $\Delta x = \Delta y = 0.1m$.

7. Appendix

A-1 Gauss-Seidel Method

Given: a two-node system as shown in Fig. 6-8a, $\Delta x = \Delta y$
Find: T1 and T2 using the Gauss-Seidel method
Sol:

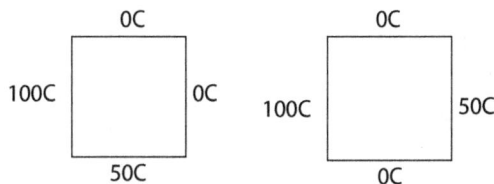

Fig. 6-7 Two 2-D square systems subject to slightly different bounday conditions

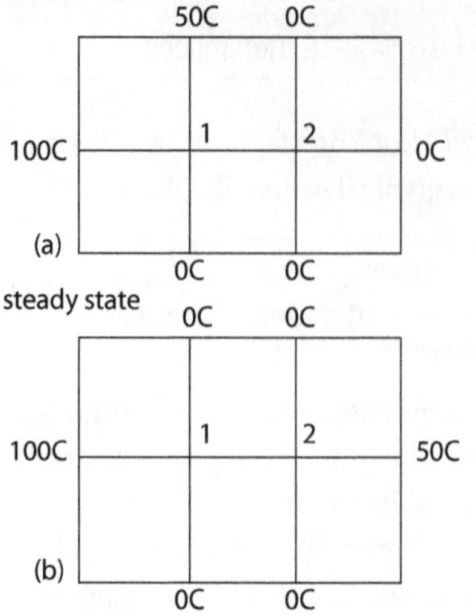

Fig. 6-8 Two rectangular grid systems subject to different boundary conditions

Using Eq. (3), we can readily obtain

T1 = 0.25*(1400 + T2), and T2 = 0.25*(1300 + T1).

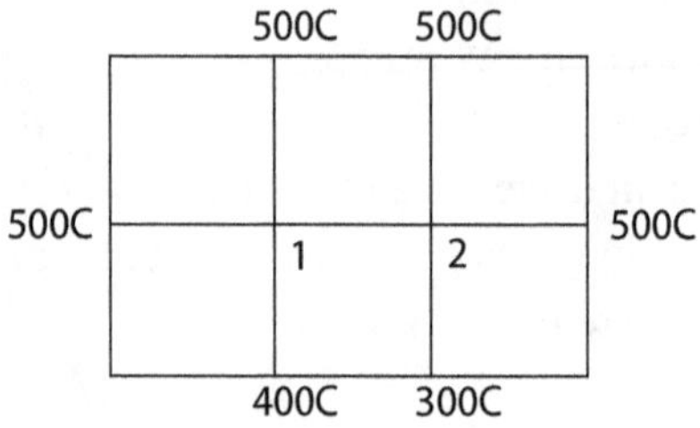

Fig. 6-8a A two-D two-node conduction system

The Matlab code solving these two equations is included below.

clc; clear
T2=500;
for iter=1:6
T1 = .25*(1400 + T2);
T2 = .25*(1300 + T1);
fprintf('%5.0f %9.2f %9.2f \n', iter, T1, T2)
end
%>>>>>>>>>>>>>>>>>>>>>>>>>>>>>

iter	T1	T2
1	475.00	443.75
2	460.94	440.23
3	460.06	440.01
4	460.00	440.00
5	460.00	440.00
6	460.00	440.00

Let us first guess T2 to be 500, thus T1 = 475 based on the governing equation for T1. Immediately, this new T1 value is substituted into the governing equation for T2, yielding T2 = 443.75. It is seen that the solution has converged to T1 = 460 and T2 = 440 at the fourth iteration.

One key procedure in the GS method is that all the terms that do not contain Ti (or Tij) should be moved to the right-hand side first. One key criterion for the GS to converge is that the coefficient matrix should be "diagonally dominant," whose definition can be found from Wikipedia [2].

The energy is conserved over a computational cell $\Delta x \Delta y$ locally. It is also conserved over the entire system globally. Let us check if the statement is true, with k = 1:

Energy_in = k* [(500 − T1) + (400 − T1) + (300 − T2)] = k*(40 − 60 − 140) = -160 W.

Energy_out = k*[(T1 − 500) + (T2 − 500) + (T2 − 500)] = k*(-40 − 60 − 60) = -160 W. Indeed, the energy is globally conserved.

A-2 A standard 2-D steady state code

```
clc; clear
k=5; qgen=1e5; % properties of the solid
Lx=0.2; nx=10; dx=Lx/nx; dxs=dx*dx; nxp=nx+1; % grid system
Ly=0.3; ny=10; dy=Ly/ny; dys=dy*dy; nyp=ny+1;
r1=(dx/dy)^2; r22=2+2*r1; qs=qgen*dxs/k; % convenient quantities
for i=1:nxp; for j=1:nyp; T(i,j)=0; end; end; % initial guess
for j=2:ny; T(1,j)=100; T(nxp,j)=0; end % boundary condition
for i=1:nxp; T(i,1)=100-(i-1)*10; end; T(:,nyp)=T(:,1);
a=zeros(nxp); b=zeros(1,nxp);
a(1,1)=1; b(1)=T(1,2); % boundary equ.
a(nxp,nxp)=1; b(nxp)=T(nxp,2); % boundary equ.
%>>>>>>>>>> compute interior nodal T's
 for iter=1:50
 for j=2:ny
 for i=2:nx
```

```
a(i,i-1)=-1; a(i,i)=2+2*r1; a(i,i+1)=-1;
b(i)=r1*(T(i,j-1)+T(i,j+1))+qs;
end
qq=a\b'; T(:,j)=qq; % update T
end
end
%>>>>>>>>>>
for i=1:nxp; for j=1:nyp
x(i,j)=(i-1)*dx; y(i,j)=(j-1)*dy;
end; end
xr=flipud(x'); yr=flipud(y'); Tr=flipud(T');
Tr
mesh(xr,yr,Tr)
% check the global energy balance
qgenT=qgen*(nx-1)*(ny-1)*dx*dy;
qW=0; qE=0; qS=0; qN=0;
for j=2:ny;
qW = qW+ k*dy*(T(1,j)-T(2,j))/dx;
qE = qE+ k*dy*(T(nx,j)-T(nxp,j))/dx;
end
for i=2:nx
qS = qS + k*dx*(T(i,1)-T(i,2))/dy;
qN = qN + k*dx*(T(i,ny)-T(i,nyp))/dy;
end
qinT = qW + qS + qgenT % = 3.105e3 W
qoutT = qE + qN % = 3.105e3 W
```

100.	90.0000	80.0000	70.0000	60.0000	50.0000	40.0000	30.0000	20.0000	10.0000	0
100.	104.6665	104.3652	100.5832	94.0749	85.2027	74.0749	60.5832	44.3652	24.6665	0
100.	112.5102	118.5625	119.3002	115.4685	107.4806	95.4685	79.3002	58.5625	32.5102	0
100.	116.8844	126.7173	130.2987	128.2135	120.8129	108.2135	90.2987	66.7173	36.8844	0
100.	119.1245	130.9378	136.0470	134.9183	127.8425	114.9183	96.0470	70.9378	39.1245	0
100.	119.8146	132.2426	137.8306	137.0039	130.0312	117.0039	97.8306	72.2426	39.8146	0
100.	119.1245	130.9378	136.0470	134.9183	127.8425	114.9183	96.0470	70.9378	39.1245	0
100.	116.8844	126.7172	130.2987	128.2135	120.8129	108.2135	90.2987	66.7172	36.8844	0
100.	112.5102	118.5624	119.3002	115.4685	107.4806	95.4685	79.3002	58.5624	32.5102	0
100.	104.6665	104.3652	100.5831	94.0749	85.2027	74.0749	60.5831	44.3652	24.6665	0
100.	90.0000	80.0000	70.0000	60.0000	50.0000	40.0000	30.0000	20.0000	10.0000	0

The 3-D plot is shown in Fig. 6-9.

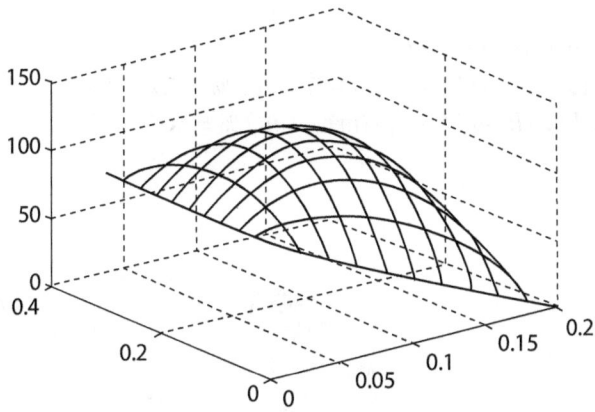

Fig. 6-9   A 3-D Matlab plot for a 2-D heat conduction problem

A-3 Optimization Problem in the Cylindrical Coordinates

```
clc; clear % cylinder
k=0.05; h=7; Tinf=-10; R1=0.001; Lx=1.76;
for iR=1:40
R2= R1 + (iR-1)*0.001;
Rplot(iR)=R2; % for plotting only
dR=R2-R1; nr=40; dr=dR/nr; nrp=nr+1;
bi=h*dr/k;
T(1)=37; T(nrp)=-10;
a=zeros(nrp); b=zeros(1,nrp);
a(1,1)=1; b(1)=T(1); %a(nrp,nrp)=1; b(nrp)=T(nrp);
a(nrp,nr)=-1; a(nrp,nrp)=1+bi; b(nrp)=bi*Tinf;
for j=1:nrp
r(j)=R1+(j-1)*dr;
end
for j=2:nr
 a(j,j-1)=r(j)-0.5*dr; a(j,j)=-2*r(j); a(j,j+1)=r(j)+0.5*dr;
end
T=a\b';
%plot(r, T);
qout(iR)=(R2-0.5*dr)*(T(nr)-T(nrp))/dr;
end
plot(Rplot, qout); grid on
xlabel('thickness of insulation'); ylabel('heat loss in W')
```

% check the global energy balance
qin = k*(2*pi*Lx)*(R1+0.5*dr)*(T(1)-T(2))/dr % = 6.7767 W
qout= h*(2*pi*Lx)*(R2-0.5*dr)*(T(nrp)-Tinf) % = 6.7767 W

Lesson 7

Lumped-Capacitance Models
(Zero-Dimension Transient Conduction)

$\mathbf{I}$n previous lessons, we studied problems that were primarily functions of x and y. This lesson marks the first time that we will formally study problems that are functions of time.

The term "zero-D" means that, in this lesson, we will assume problems to be space-independent. Sometimes, such problems are also called lumped-capacitance models, following the fact that the system is treated as a lump, whose mass, m, is usually multiplied by the heat capacity, c_v.

Nomenclature

Bi = Biot number, defined as hL/k

$c_1 = hA\Delta t/(mc_v)$

c_v = heat capacitance, J/kg-K

m = mass, kg

q''_{pan} = heat flux transferred from the frying pan to the food chunk

$r = \alpha\Delta t/(\Delta x)^2$

Thd = Thermodynamics

T^* = a parameter related to the heat flux transmitted from the frying pan, $q''_{pan}\Delta t/(\rho c_v \Delta x)$, K

α = thermal diffusivity, m^2/sec

1. Introduction

1-1 Adjectives

The phrase "functions of time" is often used interchangeably with other adjectives such as "transient," "transitory," "time-dependent," and "temporal." See the appendix for further discussions. In this textbook, we choose to use "transient" mostly.

1-2 Justifications

In this lesson, we will examine problems that are exclusively functions of time, but not space. Circumstances, in which problems, are tackled as functions of time only, include:

(a) We just desire to find some quick estimates and to see some qualitative trends, but do not care too much about accuracy.

(b) Our main interest is in the transient behavior of the system, not in the spatial distributions of temperatures and heat fluxes.

(c) The problems, indeed, are very weak functions of x and y.

1-3 Examples

Without much difficulty, we are able to list five notable transient heat conduction examples occurring in our daily life.

(a) Some people like to drink coffee in the morning. How long does it take for a cup of coffee to cool down from 70C to room temperature?

(b) How well insulated should an underground enclosure be in order for an ice block of one to melt only 10% of its mass during the time from winter to summer?

(c) During a winter snow storm, a power outage takes place in your house. Assume that there are no other heating means in the house. How long does it take for your house to cool down by 10C?

(d) Every time we open the refrigerator door, how much does this door opening add to the electricity bill, assuming that the rate is 15 cents per kW-hour?

(e) In the summer, people love to go to the beach to swim. During an hour of immersing their bodies in water at 10C, how much energy do they have to generate in order to keep their body temperature a constant 37C?

1-4 Objectives and Control Volumes

In the examples described above, even though objectives may sound different, let us try to focus first on one unknown, which is T(t) of the system. We must bear in mind that, once T(t) has been obtained, other quantities can be calculated readily.

Speaking of "the system," we must identify it in every problem clearly. In the example (a) above, the system obviously is the coffee in the cup. In (b), the system is the ice blocks in the underground enclosure. In (c, d, e), the systems are your house, the air in the refrigerator, and the human body, respectively.

The system is the control volume over which the energy conservation will be taken by using the first law of thermodynamics. It is worth noting that, once we have identified or defined the control volume, we mean to accept the assumption that the temperature inside the control volume is uniform. If T(t) should be T(t, x, y), then we must take smaller finite control volume, $\Delta x \, \Delta y \, \Delta z$, or even smaller differential control volume, dx dy dz, as shown for a 1-D case in Fig. 7-1.

1-5 The Most Important Term and the First Law of Thermodynamics

In transient heat conduction problems, the most important term is believed to be dU/dt of the system, partly because the main difference between transient problems and steady-state problems is the existence or absence of this term.

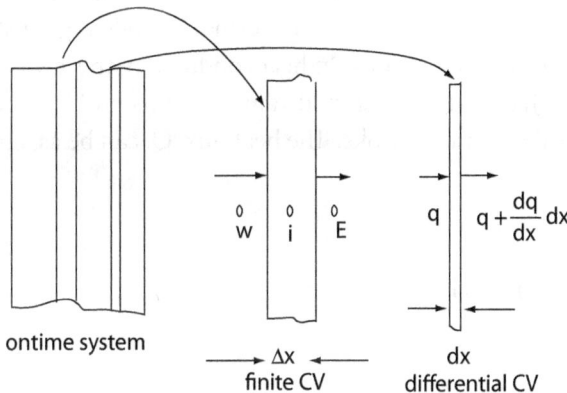

Fig. 7-1   Finding details in an energy system requires us to take small control volumes.

Usually, $\Delta U = mc_v \Delta T$. According to the first law, we obtain

$$mc_v \Delta T = (Q + W)\Delta t. \tag{1}$$

The heat fluxes, Q, take place **across** the boundaries of the control volume, whereas the work, W, in heat conduction denotes heat generation **within** the control volume. Equation (1) states that the increase of the system's internal energy is equal to the heat transfer flowing into the system and the work done on the system.

2. Detailed Analyses of a Can-of-Coke Problem

Let us start with a question that concerns every soft drink lover. How long does it take for a can of Coke, at T = 5C when taken out of the refrigerator, to warm up to 10C after it is placed on the kitchen table surrounded by room air at 25C?

2-1 Modeling

We desire to see some numbers in our answer, not just abstract rationale described by words. Seeing numbers requires establishing equations. Thus, the task that converts a problem described in words into a problem governed by equations is known as modeling. This meaning is different from the dress modeling on the stage. The task of modeling may be the most important one in the entire process of solving a heat transfer problem.

Furthermore, modeling must be based on certain laws and reasonable assumptions. In heat conduction, the law is generally nothing but Eq. (1). In Eq. (1), the term, W, should vanish for this problem, because there is no heat generation inside the can of Coke. The heat flux, Q, can be assumed to be

$$Q = hA \ (T_\infty - T). \tag{2}$$

Substituting Eq. (2) into Eq. (1) leads to

$$mc_v \ (T - T^p) = hA \ (T_\infty - T)\Delta t, \tag{3}$$

where T^p stands for the temperature of the Coke at the previous time step. Equation (3) is the governing equation that will allow us to compute T(t).

2-2 Assumptions in the Modeling

Assumptions that we have made in derivation of Eq. (3) include:

(a) T is uniform inside the Coke can,

(b) Radiation from the can to the surroundings and vice versa are neglected,

(c) During the time interval, Δt, the change of T is very small, so that T can be assumed to be constant. In passing, let us mention that

$$mc_v (T - T^p) = hA (T_\infty - T)\Delta t, \text{ known as the implicit method}$$

$$mc_v (T - T^p) = hA (T_\infty - T^p)\Delta t, \text{ known as the explicit method}$$

$$mc_v (T - T^p) = 0.5hA (T_\infty - T)\Delta t + 0.5hA (T_\infty - T^p)\Delta t, \text{ known as the Crank-Nicolson method [1]}$$

Each method is associated with its own merit. The implicit method is stable. The explicit is quick and easy because the information on the right-hand side of the equation is all known. Finally, the Crank-Nicolson method is accurate. We will adopt the implicit method most of the time in this textbook.

Equation (3) can be solved logically and numerically by writing a Matlab code and running it. In the next section, we will present the code with explanations. Before coding, let us rearrange Eq. (3) into

$$T = (T^p + c_1 T_\infty)/(1 + c_1), \tag{4}$$

where $c_1 = hA \Delta t/(mc_v)$.

2-3 A Matlab Code Computing T(t) for the Can of Coke Problem

```
clc; clear
m=.28; cv=4180; h=4.8; A=.01; % properties of the Coke
Tinf=25; Tp=5; % information regarding the temperatures
time(1)=0; T(1)=Tp; % initial condition
Lt=720*60; % the total length of time is 12 hours
nt=200; % number of time intervals
dt=Lt/nt; % the length of the time interval
c1=h*A*dt/(m*cv); % a convenient dimensionless parameter
for it=2:nt+1
time(it)=(it-1)*dt; % for example, time(3)=2*dt
T(it)= (Tp+c1*Tinf)/(1+c1); % governing equation
Tp=T(it); % must remember to update Tp
```

end
plot(time/3600, T); grid on;
xlabel('time in hours'); ylabel('T of Coke')
times=interp1(T, time, 10)/60 % = 117.43 minutes
*text(times/60-0.1, 10, '**')*

According to Fig. 7-2, it takes 117.43 minutes for the Coke to warm up from 5C to 10C.

2-4 Discussions

A few comments on the computation can be made below:

(a) During the first 6 hours of warming, the Coke system has gained about 12C. During the next 6 hours, it has warmed up only by 5C. This phenomenon occurs in daily life often. If we would like to take upon the hobby of piano playing today, for example, we may experience rapid improvement, prompting us to think that we are talented. After 6 months, however, after we have learned to play "Twinkle Twinkle Little Star" with both hands, we start improving slowly and feeling discouraged.

(b) The heat capacity of water, cv, is relatively high. If we replace water with other types of fluids, we will see the curve shift up, suggesting that the system will warm up more rapidly.

(c) If we increase the mass of the drink, the curve will shift down. Therefore, it takes longer for a 2-liter bottle of Coke to warm up.

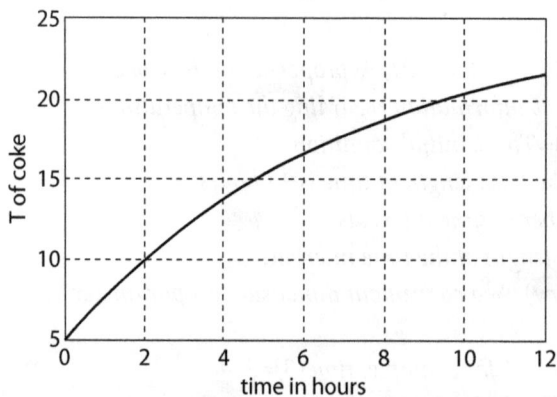

Fig. 7-2   The temperature of the coke drinks as a function of time

(d) It is a good habit to check if the total energy gained by the Coke, ΔU is equal to the total heat transfer entering the Coke from the warm air, Q_{in}. Thus, let us add a few Matlab coding lines as:

```
% total energy balance
dU= m*cv*(T(201)-T(1)) % = 1.9396e4 J
sum = 0;
for it=2:nt+1
 sum = sum + h*A*(Tinf - T(it))*dt;
end
 Q_in= sum % = 1.9396e4 J
```

Indeed, it is pleasantly found that these two terms are equal. Hence we have more confidence in the bug-free status of the code.

(e) A quick comparison between T in the Lumped-Capacitance Model and T_boundary in the heat conduction problems reveals that their governing equations resemble each other. They are, respectively,

$$T = (T^p + c_1 \, T_\infty)/(1 + c_1), \text{ from Eq. (4)}$$

$$T_{11} = (T_{10} + bi \, T_\infty)/(1 + bi), \text{ if node 11 is on the boundary.}$$

These two equations will reduce to $1 = 1$ if T has reached steady state, and if T_{11}, T_{10}, and T_∞ have reached thermal equilibrium.

3. When Is It Appropriate to Use the Lumped-Capacitance Model?

In the literature, it has been recommended that a thermal system can be safely assumed to be a lumped-capacitance model if the Biot number, defined as, hL/k, is less than 0.002. Let us use a coarse three-node model to examine two cases below.

3-1 A Three-Node Example

Example 7-1.

Consider two vertical slabs, one of which is made of aluminum and the other wood. Relevant data are h = 100 W/m^2-K, T_∞ = 100C, L = 0.01 m, and Δt = 20 sec. Initially, the two slabs all are at 0C. Suddenly, they are exposed to the hot airflow. See Fig. 7-3 for the system schematic.

Find: T1, T2, and T3 during the first two time steps.

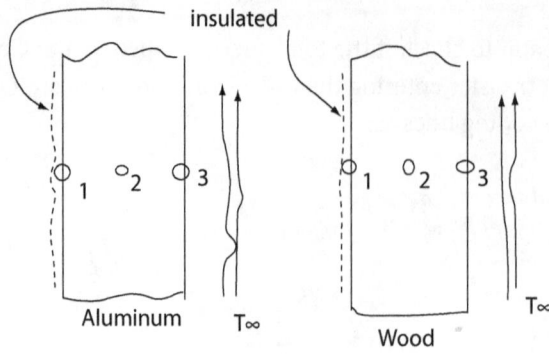

Fig. 7-3 Conduction within two systems. The lumped-capacitance model should not be applied to the case of wood slab.

Sol: The governing equations for T1, T2, and T3 can be derived as:

T1 = T2, due to insulation at x = 0,

T2 = T2p + r*(T1 - 2*T2 + T3), its derivation will be explained in lesson 8,

T3 = (T2 + bi*)/(1+bi).

The solutions are listed below

aluminum, bi = 0.0021, r = 77.6 ---- LC model is acceptable

t	T1	T2	T3
0	0	0	0
20	0	0	0.2105
40	16.337	16.337	16.513 ---- fairly uniform

wood,bi = 2.94, r = 0.104 -------- LC model should not be used.

t	T1	T2	T3
0	0	0	0
20	0	0	74.62
40	7.76	7.76	76.596 ----- quite non-uniform

3-2 An Analogy

As an analogy, a daily life example easier for us to remember the Biot-number restriction, let us imagine the right face to be a seashore. A large *B*iot number means that the number of *b*oats is large. Similarly, large thermal conductivity, *k*, means that the number of truc*k*s is large.

When there are many boats, many cargoes (energy) are shipped to the coast, resulting in huge surpluses accumulated in the storage rooms (high temperature at x=L). However, if there is an equal number of trucks to ship these cargoes inland, then the cargo distribution will tend to be more uniform.

4. A Simple Way to Relax the Bi < 0.002 Constraint

A simple way to raise the Biot-number restriction is given below. The system is shown in Fig. 7-4. Let us assume

$$T(\xi) = T_1 + d_1\xi + d_2\xi^2, \text{ where } \xi = x / L. \tag{5}$$

We can work on T(x) instead. Using ξ is just the author's preference. There are four unknowns: T_1, T_m, d_1, and d_2. Their associated conditions can be found below:

We can try hard by algebraically manipulating to solve the four equations analytically if we like. But, again, our preference may be to enslave the computer.

At $\xi=0$, $h(T_\infty - T_1) = -k \left(\dfrac{dT}{dx}\right)_{x=0}$, or

$$\text{Bi} *(T_\infty - T_1) = -d_1. \tag{6a}$$

At $\xi=1$, due to insulation at x=L,

$$d1 + 2\, d2 = 0, \tag{6b}$$

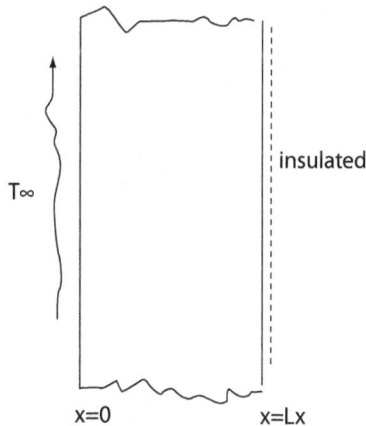

Fig. 7-4   **A lumped-capacitance system, with the right face behaving as the line of symmetry, or equivalently an insulated plane**

According to the definition of T_m, namely, $T_m = \int_0^1 T(\xi)d\xi$, we obtain

$$T_m = T_1 + \frac{d_1}{2} + \frac{d_2}{3} \tag{6c}$$

Global energy balance leads to: $h \, \Delta t \, (T_\square - T_1) = \rho c_v L_x (T_m - T_m^p)$, or

$$C_1(T_\infty - T_1) = T_m - T_m^p \tag{6d}$$

In the Matlab code, let q(1) = T1, q(2) = Tm, q(3) = d1, and q(4) = d2.

Equations (6a, b, c, d) allow us to deviate from the lumped-capacitance model, in which $T_m = T_1 = T$ everywhere. This treatment has lifted up the Biot-number restriction to approximately. See the Matlab codes and results in the appendix.

5. Why Stirring the Food When We Fry It?

When we fry food, such as broccoli, why do we stir it around? It seems so intuitive that we should do so. But let us find some numbers to support our intuition below.

Figure 7-4a shows two chunks of food piling up on the frying pan schematically. If we adopt the lumped-capacitance model, then T1 and T2 can represent the uniform temperatures inside chunk1 and chunk2, respectively. At a given instant, the equation governing T1 and T2 can be written as

$$mc_v \, (T_1 - T_1^p)/\Delta t = Aq_{pan}'' - Ak(T_1 - T_2)/\Delta x \tag{7a}$$

and

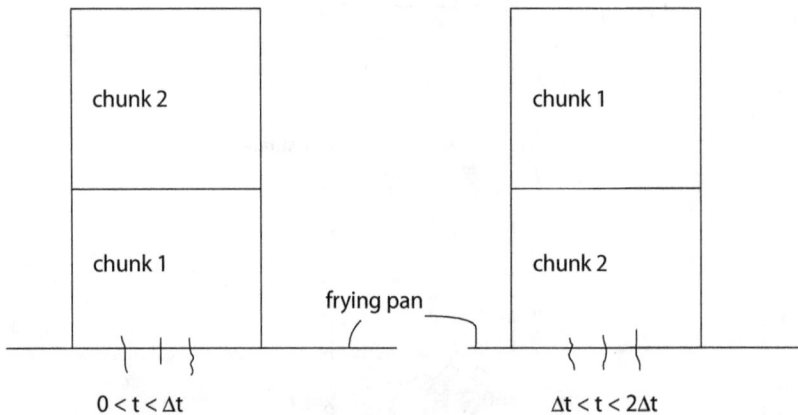

chunk 2	chunk 1
chunk 1	chunk 2

frying pan

$0 < t < \Delta t$ $\Delta t < t < 2\Delta t$

Fig. 7-4a   **Stirring two chunks of broccoli on the frying pan. At the end of the first time step, the two chunks are turned over.**

$$mc_v (T_2 - T_2^p)/\Delta t = Ak(T_1 - T_2)/\Delta x \tag{7b}$$

Let us assume that these two chunks of food are cubes. Thus,

$$m = \rho V = \rho(\Delta x)^3.$$

Equations (7a, b) can be simplified into

$$(1+r) T_1 - rT_2 = T_1^p + T^*, \tag{8a}$$

$$-rT_1 + (1 + r) T_2 = T_2^p, \tag{8b}$$

where r and T^* are defined in the nomenclature.

The first code computes the case of stirring. The second code computes the case of no stirring at all. It can be seen that the former case yields a much more uniform temperature, which is desired by us.

```
clc; clear % with stirring
k=0.61; rho=1000; cv=4180; aLf=k/(rho*cv);
q_fry=2e4;
dx=0.003; dxs=dx*dx;
dt=10; r=dt*aLf/dxs; Ts=q_fry*dt/(rho*cv*dx);
Tp(1)=20; Tp(2)=20;
a=[1+r -r; -r 1+r]; b=[Tp(1)+Ts Tp(2)];
qq=a\b'; qq'
Tp(1)=qq(2); Tp(2)=qq(1); % update Tp. Two chunks are switched.
% 2nd time step
b=[Tp(1)+Ts Tp(2)];
qq=a\b'; qq' % = 37.42C 34.4744C
%%>>>>>>>>>>>>>>>>>>>>>>>>>>>>>>>>>>>>>>
clc; clear % without stirring
k=0.61; rho=1000; cv=4180; aLf=k/(rho*cv);
q_fry=2e4;
dx=0.003; dxs=dx*dx;
dt=10; r=dt*aLf/dxs; Ts=q_fry*dt/(rho*cv*dx);
Tp(1)=20; Tp(2)=20;
a=[1+r -r; -r 1+r]; b=[Tp(1)+Ts Tp(2)];
qq=a\b'; qq'
Tp(1)=qq(1); Tp(2)=qq(2); % update Tp
% 2nd time step
b=[Tp(1)+Ts Tp(2)];
qq=a\b'; qq' % = 46.52C 25.38C
```

6. Summary

In this lesson, we have learned how to solve the simplest problem in transient heat conduction. Three topics are studied:
   (a) warming of the Coke in a can
   (b) the restriction of using the lumped-capacitance model
   (c) relaxing the restriction stated in (b).

7. Reference

1. J. Crank and P. Nicolson, *A Practical Method for Numerical Evaluation of Solutions of Partial Differential Equations of the Heat Conduction Type*, Proc. Camb. Phil. Soc. 43: 50–67, 1947.

8. Exercise Problems

8-1 A vertical energy-generating chip is shown in Fig. 7-4. Its right face is insulated. Its left face is exposed to a hot air flow at 100C. Let us assume that k is sufficiently large such that the system can be assumed to be a lumped-capacitance model. All the input data are given below in the Matlab code. Initially, the chip is at T=0C; h=130; rho=1000; cv=2300; q_g=2.5e5.

As L increases, the total amount of heat generation increases. Hence T(at t = 1000 sec) increases. However, when L increases to a certain thickness, it is possible that, because the mass also increases, the total heat capacitance increases as well. The latter effect dominates the former effect, and starts to render the system to be heated slowly. Therefore, beyond this L_critical, as L increases, T(t = 1000 sec) may decrease.

The governing equation for T_chip is given as:

$$\rho c_v L \frac{\partial T}{\partial t} = h(T_\infty - T) + L q_{gen}'''$$

Determine L_critical.

8-2 The temperature history, T(t), of a certain lumped-capacitance system in a cold air stream is shown in Fig. 7-5. If c_v increases, should the curve

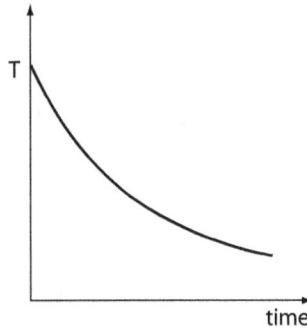

Fig. 7-5   A typical history of T(t) for a lumped-capacitance system

(a) shift up, (b) shift down, (c) remain the same,
(d) first shift up, then shift down?

8-3 The kitchen is a good place for us to learn heat transfer. According to your mom's experience, T_egg reaches 97.5C after 5 minutes of boiling the water in the pot. Let us assume egg is a lumped-capacitance system.

The characteristic of this problem is that, in addition to T_egg = function of time, T_inf (or T_water) is also a function of time. All the data are given as.

rho=1000 kg/m^3; cv=4180 J/kg-K; R_egg = 0.015m
(assume that eggs are spheres);

T_egg_initial = 5C; Δt = 6 sec; the heat transfer coefficient, h, initially is 100 W/m^2-K, increasing by 30 W/m^2-K at every time step.

$T_\infty(t)=20+0.5333*t - 8.8889e\text{-}4*t^2;$

Find: h

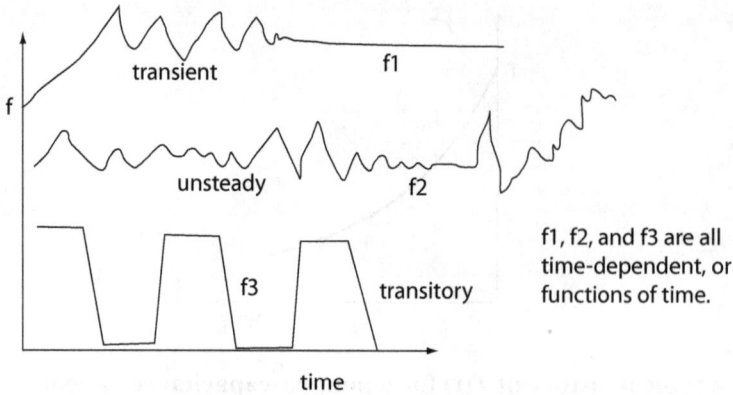

Fig. 7-6 A graph depicting meanings of 3 similar adjectives

9. Appendix

A-1 Discussions of Adjectives

There are six phrases or words presented here that are related to time. Are they really interchangeable, or are people simply using them non-rigorously, or perhaps both? Discussions here are meant to be for a laugh more than being serious, and are only the author's own view.

(a) "Functions of time" may be interchangeable with "time-dependent."
(b) "Temporal" means "related to time." We may say, "dT/dt" is a temporal term, as opposed to dT/dx, which is a space-related term.
(c) "Unsteady," "transient," and "transitory" can be best distinguished by sketching them as shown in the figure.

In reference to Fig. 7-6, three curves, f1, f2, and f3, look somewhat different. Curve f1 is expected to eventually reach steady state; curve f2 is expected to be unsteady forever; curve f3 focuses on its "on and off" or "live or die" nature. For example, it is better for us to say:

Everything on Earth is transitory; the only thing or attribute that is not transitory is "transitory" itself.

A-2 Comparisons of Three Cases:

```
%>>>> quadratic approximation
clc; clear
h=200; Tinf = 100; Lx=0.2; rho=1000; cv=2300; k=20; dt=20;
```

```
Tmq(1)=0; time(1)=0; Tmp=Tmq(1);
c1=h*dt/(rho*cv*Lx); Bi=h*Lx/k;
for itm = 2:601
 time(itm)=(itm-1)*dt;
 a=[-Bi 0 1 0; 0 0 1 2; 1 -1 1/2 1/3; c1 1 0 0];
 b=[-Bi*Tinf 0 0 c1*Tinf+Tmp];
 q=a\b'; Tmp=q(2); Tmq(itm)=q(2);
end
%
%>>>> 1-D transient conduction
clear a b
nx=40; nxp=nx+1; dx=Lx/nx; dxs=dx*dx; aLf=k/(rho*cv);
r=aLf*dt/dxs; Tma(1)=0; bi=h*dx/k;
a(nxp,nxp)=0; b(nxp)=0; Tp(nxp)=0;
for itm =2:601
sum =0;
a(1,1)=1+bi; a(1,2)=-1; b(1)= bi*Tinf;
for i=2:nx
 a(i,i-1)=-r; a(i,i)=2*r+1; a(i,i+1)=-r; b(i)=Tp(i);
end
a(nxp,nx)=-1; a(nxp, nxp)=1;
T=a\b'; Tp=T;
for i=1:nxp
 sum = sum +T(i);
end
Tma(itm)=sum/nxp;
end
%>>>> lumped-capacitance model
Tmz(1)= 0; Tmp=0;
for itm=2:601
 Tmz(itm)= (Tmp + c1*Tinf)/(1+c1); Tmp=Tmz(itm);
end
%>>>> Final Plot
plot(time, Tmq,'k--', time, Tma,'r', time, Tmz,'k-.')
xlabel('time in sec'); ylabel('T-mean');
text(4000,60,'solid:1-D'); text(4000, 53,'dash line: quadratic');
text(4000,46,'dash-dot line: lumped-capacitance')
```

Lesson 8

One-Dimensional Transient Heat Conduction

L et us continue studying the subject of transient heat conduction. In this lesson, the focus will be on finding the solution of T(t, x).

Nomenclature

A_c = cross-sectional area of the fin, m^2

$c_1 = (hp\Delta t)/(\rho c_v A_c)$

D = denoting "dimension"

m = the mass of the control volume, kg

p = perimeter of the fin, m

r = aspect ratio, defined as $\alpha\Delta t/(\Delta x)^2$

T^p = temperature at the previous time step, C

1. Kitchen Is a Good Place to Learn Heat Transfer

Not only is a kitchen a very important place feeding our stomachs, but also it is actually a good technical source that is capable of teaching us heat transfer.

You are such a diligent student who has been studying heat transfer hard all day. Your mom is about to fry some potato patties to reward you. Based on the heat conduction knowledge you have learned, do you think you are able to advise her on how she can avoid burning the patties?

1-1 Governing Equation of One-D Transient Heat Conduction

A patty is vertically drawn in Fig 8-1. Its left face is subject to the heat flux generated from the frying pan. If we are interested in the solution of T(t, x), we need to take energy balance over a small segment, Δx, inside the patty. Applying $\Delta U = Q$ to this segment, we obtain

$$mc_v\,(T_i - T_i^p) = (q_w - q_e)\Delta t\,, \tag{1}$$

where nodes denoted by lowercase letters, w and e, are located on the boundary of the segment Δx.

It is an important issue to pay attention to how we subdivide and index the patty system. See the grid in Fig. 8-1. Such a grid convention will be adopted throughout this textbook. So the boundaries of the control volume are positioned halfway between the nodes.

In Eq. (1), we further have

$$m = \rho A \Delta x\,,\ q_w = kA\,(T_W - T_i)/\Delta x,\ q_e = kA\,(T_i - T_E)/\Delta x,$$

which can be substituted into Eq. (1) to yield, after straightforward algebra,

$$-rT_W + (2r + 1)T_i - rT_E = T_i^p. \tag{2}$$

Equation (2) can be applied to all the interior nodes and thus serves as the most important piece of information in the subject of one-D transient heat conduction problems. On the boundary, however, it should not be used. Temperatures on the two

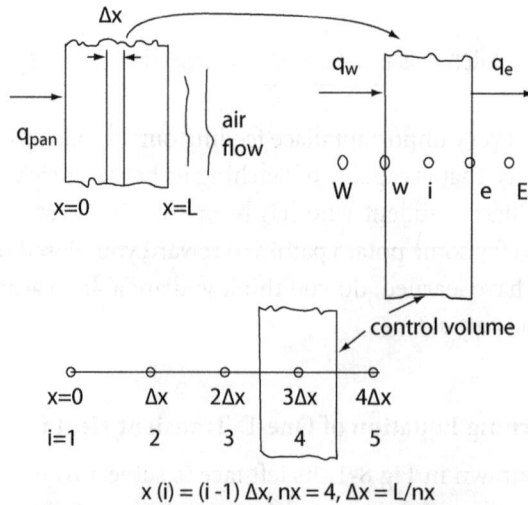

Fig. 8-1   A 1-D transient heat conduction system, its computational cell, and the relationship between x and i.

boundaries, x=0 and x=L, should be either given or derived by using the sheet energy balance.

See Problem 8-1.

1-2 The Generic Core of One-D Transient Code

Independent of any types of boundary conditions, Eq. (2) can be coded conveniently as follows:

for i = 2:nx
a(i,i-1)= -r; a(i,i)= 2*r + 1; a(i,i+1)= -r; b(i)= Tp(i);
end

This simple *for loop* provides us with nx-1 equations for all the interior unknowns. There can be all sorts of activities, such as radiation, two-phase flows, or Halloween parties taking place on the boundaries. But they should not affect Eq. (2) or this for loop directly.

The indices, i=1 and i=nx+1, are reserved for T(1) and T(nx+1).

1-3 Various Boundary Conditions

Let us now make a table to show all the algebraic equations associated with three possible types of boundary conditions at i=1 and i=nx+1 in Table 8-1 below. Notice that none of the equations contains the term T_i^p, because a sheet does not have any capacity to store any energy. If energy cannot be stored, T_i^p should not appear.

Table 8-1 *Governing equations for boundary nodal temperatures T(1) and T(nx+1)*

BC type	*given*	*i = 1*	*i = nxp (nx + 1)*
Dirichlet	Tn	a(1,1)=1; b(1)=Tn;	a(nxp,nxp)=1; b(nxp)=Tn;
Neumann	q''	a(1,1)=1; a(1,2)=-1; b(1)=qb;	a(nxp,nx)=1; a(nxp,nxp)= -1; b(nxp)=qb;
Mixed	h	a(1,1)=1+bi; a(1,2)=-1; b(1)=bi*T_∞;	a(nxp,nx)=-1; a(nxp,nxp)=1+bi; b(nxp)=bi*T_∞

where qb = $q''\Delta x/k$ and bi = $h\Delta x/k$.

1-4 A One-D Transient Matlab Code

The input data are briefly described:
the heat flux from the frying pan to the patty = 1200 W/m^2,

the heat transfer coefficient, h, between the top of the patty and the kitchen air = 4.5 W/ m^2—K

the total length of time spent to fry the patties = 1800 sec

the initial temperature of the patties = -2C

the temperature of the kitchen air = 25C

For this patty-frying problem, at i=1, the Neumann boundary condition is used; at i=nxp, the mixed-type boundary condition is used. Always remember to update our Tp(i) as we march along in time.

```
clc; clear
%>>>>>>>>>> input data
qs=1200; h=4.5; Tinf=25;
k=0.2; rho=800; cv=3000; aLf=k/(rho*cv);
L=0.009; nx=10; dx=L/nx; dxs=dx*dx; nxp=nx+1;
Lt=1800; nt=20; dt=Lt/nt;
r=aLf*dt/dxs; bi=h*dx/k; qb=qs*dx/k;
x=linspace(0,L,nxp);
%>>>>>>>>>> initialize a and b, and prescribe the B. C.'s
a(nxp,nxp)=0; b(nxp)=0;
a(1,1)=1; a(1,2)=-1; b(1)=qb;
a(nxp,nx)=-1; a(nxp,nxp)=1+bi; b(nxp)=bi*Tinf;
%
%>>>>>>>>>> initial condition
for i=1:nxp
Tp(i)=-2;
end
T1(1)=-2; time(1)=0;
%
%>>>>>>>>>> interior-T gov. eq.
for i = 2:nx
a(i,i-1)= -r; a(i,i)= 2*r + 1; a(i,i+1)= -r;
end
for it=2:nt+1
time(it)=(it-1)*dt;
for i=2:nx; b(i)= Tp(i);end
T=a\b; Tp=T;
T1(it)=T(1); Tnxp(it)=T(nxp); % = store up values of T1 and Tnxp
plot(x,T); hold on; xlabel('x'); ylabel('T')
text(0.001, 130, 'As time increases, the curve shifts up')
```

```
end
%plot(time, T1)
hold off
%>>>>>>>>>> check the total energy balance
sum1=0; sum2=0;
for it=2:nt+1
sum1 =sum1+(qs-h*(Tnxp(it)-Tinf))*dt;
end
for i=2:nx
 sum2=sum2 + dx*rho*cv*(T(i)-(-2));
end
q_total = sum1 % = 1.9901e6 J/m^2
dU=sum2 % = 1.9901e6 J/m^2
```

If some readers find this *for loop* too abstract to understand, please try to first work on a coarse grid of three or four nodes for only the first time step without having to use the space *for loop* and the time *for loop*. By calculating only three or four unknowns at only one time step, we can use the calculator to do the job. See Fig. 8-2 for the temperature distributions changing in time.

See Problem 8-2.

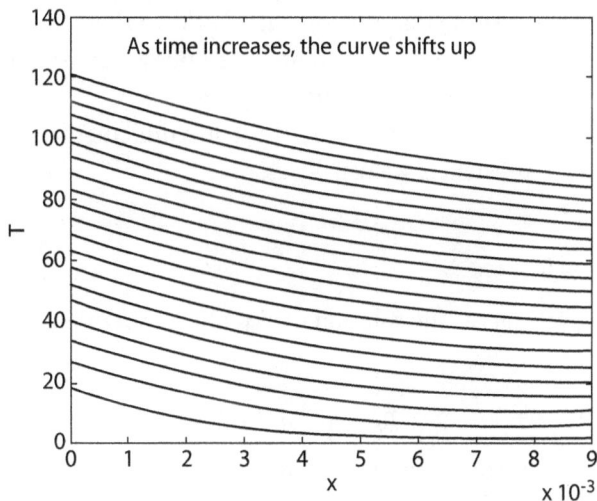

Fig. 8-2   **T of the potato patty as a function of x, parameterized in time**

1-5 Global Energy Balance

After we have written and run the Matlab code, it is often possible and beneficial us to check the validity of the code. One reliable way to do the checking is to find if the energy is conserved globally over the entire period of time. In this patty-frying problem, we should have:

internal energy increase of the patty = total heat flow entering the patties – total heat flow escaping from the patties to the kitchen air. (3)

Based on our computations, LHS of Eq. (3) is identical to RHS of Eq. (3), suggesting that the code is bug-free.

2. Other One-D Transient Heat Conduction Applications

Listed below are a few examples related to one-D transient heat conduction. Their respective Matlab codes are included in the appendix.

2-1. Semi-Infinite Solid

Consider the earth ground initially at 20C. Suddenly, a cold front arrives, so that the surface temperature becomes -10C, and is kept at -10C all the time. We are interested in finding T(t, x) of the soil during a month.

In this problem, the boundary condition at x=0 is obvious, and is T(1) = -10C. At x=L, we do not know clearly what will be going on. If we take L to be one mile deep, we probably can safely assume that the soil particle down there will not be affected by the arrival of the cold front within a month. But taking our computational domain unnecessarily too largely will waste computation time.

On the other hand, if L is taken to be too short, it is difficult to specify the boundary at x=L. Under this circumstance, we usually assume

$\partial T/\partial x \approx 0$ at $x = L$,

where L is relatively large. The solution of T(t, x) is given in Fig. 8-3.

2-2. Revisit Fin Problems

A one-D fin is different from a one-D slab in that in the former, there is an additional term of heat transfer via the circumferential area of the fin. Therefore, we should re-derive the governing equation for T(i) of a fin, and should not use Eq. (2) carelessly.

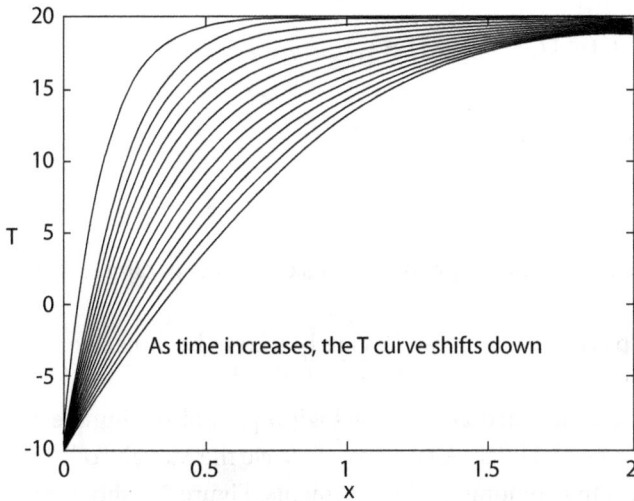

Fig. 8-3   Temperature distributions changing in time inside a semi-infinite solid

Energy balance over the control volume i, shown in Fig. 8-4, leads to:

$$\Delta U = (q1 - q2 - q3)\Delta t,$$

where

$$q1 = kA_c(T_{i-1} - T_i)/\Delta x,$$
$$q2 = kA_c(T_i - T_{i+1})/\Delta x,$$
$$q3 = hp\Delta x\,(T_i - T_\infty),\ \text{and}$$
$$\Delta U = \rho A_c \Delta x c_v\,(T_i - T_i^p).$$

After substitution and simplification, we can obtain

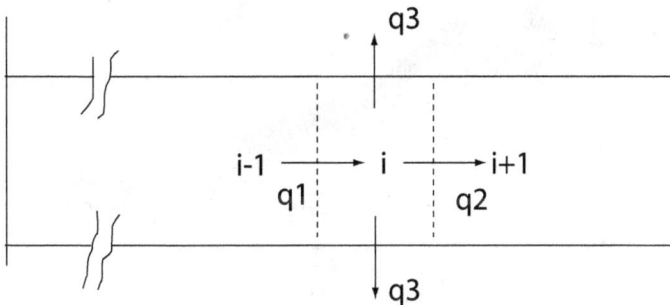

Fig. 8-4   Transient energy balance over a control volume of the fin

$$-rT_{i-1} + (2r + 1 + c_1)T_i - rT_{i+1} = T_i^p + c_1 T_\infty,$$ (4)

where

$$c_1 = (hp\Delta t)/(\rho c_v A_c).$$

If we like, we can further express in terms of known parameters as:

$c_1 = \text{bi} * r * (p \,\Delta x/A_c)$, which can be clearly seen to be dimensionless.

Once we have established Eq. (4), the logical path of solving the problem is similar to that of solving a one-D slab. Again, we will leave the Matlab code in the appendix for interested readers to run numerical experiments. Figure 8-5 shows the solution T(t, x).

Post-processing curiosity may include:

(1) varying values of k, ρ, c_v, h, D and L, observing the shifting trend of temperature profiles,
(2) plotting q'' versus x, parameterized in time, and varying parameters to see trends,
(3) solving an optimization problem with the constraint being fixed volume of the fin, and the objective being to find the maximum heat loss rate within 10 minutes. See Problem 8-4.

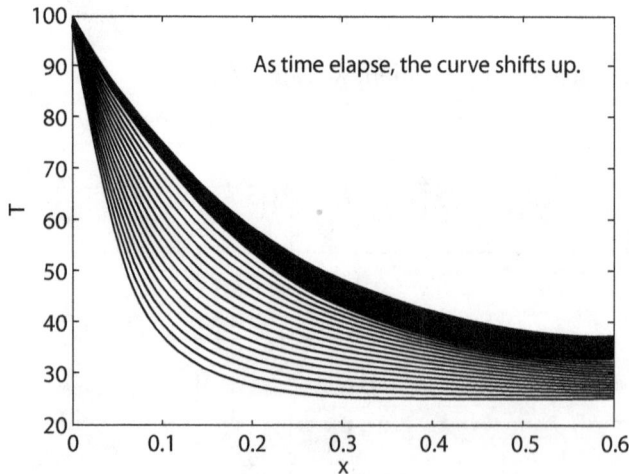

Fig. 8-5   T of a 1-D fin as a function of x parameterized in time

2-3. Multi-Layer Slabs with Heat Generation

Revisit Example 4.1 regarding 1-D steady-state heat conduction within a system of multi-slabs with nonzero heat generation. Suddenly, slab B is removed. All other conditions remain the same. If we are interested in finding $T_A(t,x)$, then the new problem belongs to the subject of 1-D transient heat conduction.

The characteristic of this problem is that the initial condition is not uniform temperature, as usually given in other problems, which we need to pay attention to.

See also Problem 8-1.

3. Differential Governing Equation for One-D Transient Heat Conduction

Between Eq. (1) and Eq. (2), an intermediate step can be written as

$$T_i - T_i^p = \frac{\alpha \Delta t}{(\Delta x)^2} (T_W - 2T_i + T_E).$$

Letting Δt and Δx shrink to infinitesimal sizes, we will obtain

$$\frac{\partial T}{\partial t} = a \frac{\partial^2 T}{\partial x^2}, \tag{5}$$

which is a classical one-D transient heat conduction governing differential equation.

4. Summary

This lesson provides us with basic understanding of finding T(t, x). Important topics include:

(a) one-D transient heat conduction of potato-patty frying,
(b) knowing different types of boundary conditions,
(c) a good habit to check the energy balance globally (in space) and totally (in time)
(d) some one-D transient heat conduction applications

5. References

(1) Wikipedia for thermal properties of potato patties

8-1 Modify Eq. (2) if there is heat generation in the slab.

8-2 Judging from the figure of $T(x, t)$ for the potato-patty frying problem, we realize that the system can be almost treated as a lumped-capacitance model. Proceed to treat it so, find $T(t)$, and compare $T_m(t)$ computed from the official transient 1-D code.

8-3 At time = 10 minutes, check the global energy balance for the fin.

8-4 Consider a fin with both ends insulated. The diameter of the fin becomes large, so that the temperature variations in the r direction can no longer be neglected. Derive the algebraic governing equation for $T(t, r)$.

7. Appendix

A-1 Semi-Infinite Solids

```
clc; clear
%>>>>>>>>>> input data, semi-infinite solid
k=0.5; rho=2023; cv=1825; aLf=k/(rho*cv);
L=2; nx=60; nxp=nx+1; dx=L/nx; dxs=dx*dx;
Lt=30*24*3600; nt=20; dt=Lt/nt;
r=aLf*dt/dxs;
T_coldfront=-10; T_initial=20;
x=linspace(0,L,nxp);
%>>>>>>>>>> initial condition or initializing
a=zeros(nxp,nxp); b=zeros(1,nxp);
Tp=T_initial*ones(1,nxp); time(1)=0;
%>>>>>>>>>> prescribe the B. C.s
a(1,1)=1; b(1)=T_coldfront;
a(nxp,nx)=-1; a(nxp,nxp)=1; % assume dT/dx=0 at x=L
%>>>>>>>>>> interior-T gov. eq.
for i = 2:nx
a(i,i-1)= -r; a(i,i)= 2*r + 1; a(i,i+1)= -r;
end
for it=2:nt+1
time(it)=(it-1)*dt;
for i=2:nx; b(i)= Tp(i);end
T=a\b'; Tp=T;
plot(x,T); hold on; xlabel('x'); ylabel('T')
```

```
end
axis([0,2,T_coldfront,T_initial]);
text(0.3, -3, 'As time increases, the T curve shifts down')
hold off
% check the energy balance globally and totally
%
sum1=0; sum2=0;
for it=2:nt+1
sum1 =sum1+(k*(T(1)-T2(it))/dx)*dt;
end
for i=2:nx
 sum2=sum2 + dx*rho*cv*(T(i)-(T_initial));
end
q_total = sum1 % = -7.1754e7 J/m^2
dU=sum2 % =% -7.1754e7 J/m^2
%>>>>>>>>>>
```

A-2 Transient 1-D Fins

```
clc; clear
%>>>>>>>>>> input data
h=10; Tinf=25; Tb=100;
k=398; rho=8933; cv=385; aLf=k/(rho*cv);
D=0.006; Ac=D*D*pi/4; p=pi*D;
L=0.6; nx=20; dx=L/nx; dxs=dx*dx; nxp=nx+1;
Lt=1200; nt=40; dt=Lt/nt;
r=aLf*dt/dxs; c2=(h*p/k)*(aLf*dt/Ac);
x=linspace(0,L,nxp);
%>>>>>>>>>> initialize a and b, and prescribe the B. C.'s
a(nxp,nxp)=0; b(nxp)=0;
a(1,1)=1; b(1)=Tb;
a(nxp,nx)=-1; a(nxp,nxp)=1;
%
%>>>>>>>>>> initial condition
for i=1:nxp
Tp(i)=Tinf;
end
time(1)=0;
%>>>>>>>>>> interior-T gov. eq.
for i = 2:nx
```

```
a(i,i-1)= -r; a(i,i)= 2*r+1+c2; a(i,i+1)= -r;
end
qconv(nt+1)=0;
for it=2:nt+1
time(it)=(it-1)*dt;
for i=2:nx; b(i)= Tp(i)+c2*Tinf; end
T=a\b'; Tp=T;
T2(it)=T(2); % = store up values of T2
plot(x,T); hold on; xlabel('x'); ylabel('T')
for i=2:nx
qconv(it) = qconv(it) + dx*p*h*(T(i)-Tinf);
end
end
hold off
text(0.2, 90, 'As time elapses, the curve shifts up.')
axis([0,L,20,Tb]);
%>>>>>>>>>> check the total energy balance
sum1=0; sum2=0;
for it=2:nt+1
qbase=k*Ac*(Tb-T2(it))/dx;
sum1 =sum1+(qbase-qconv(it))*dt;
end
for i=2:nx
 sum2=sum2 + dx*Ac*rho*cv*(T(i)-Tinf);
end
q_total = sum1 % = 1.6038e3 J
dU=sum2 % = 1.6038e3 J
```

A-3 Example 4-1 Revisited

```
clc; clear
ka=75; La=.05; Qa=1.5e6;
kb=150; Lb=.02;
h=1000; T_inf=30;
nx=300; nxp=nx+1; nxq=nx+2; dx=La/nx; dxs=dx*dx;
c1=Qa*dxs/ka; c2=(kb/ka)*(dx/Lb); Bi=h*Lb/kb;
a=zeros(nxq); b=zeros(1,nxq);
x1=linspace(0,La,nxp); x=[x1, La+Lb];
a(1,1)=1; a(1,2)=-1;
for i=2:nx
```

```
a(i,i-1)=1; a(i,i)=-2; a(i,i+1)=1; b(i)=-c1;
end
a(nxp,nx)=1; a(nxp,nxp)=-(1+c2); a(nxp,nxq)=c2;
a(nxq,nxp)=1; a(nxq,nxq)=-(1+Bi); b(nxq)=-Bi*T_inf;
T=a\b';
%>>>>> suddenly, material B is removed.
rho=8000; cv=450; dtm=300;
Qgs=Qa*dtm/(rho*cv); aLf=ka/(rho*cv); r=aLf*dtm/dxs;
bi=h*dx/ka;
Tp=T(1:nxp);
plot(x1, Tp); hold on % initial condition
for itm=2:11
aa(1,1)=1; aa(1,2)=-1;
for i=2:nx
aa(i,i-1)=-r; aa(i,i)=1+2*r; aa(i,i+1)=-r; bb(i)=Tp(i)+Qgs;
end
aa(nxp,nx)=1; aa(nxp,nxp)=-(1+bi); bb(nxp)=-bi*T_inf;
Ta=aa\bb'; Tp=Ta;
plot(x1,Ta); hold on
end; hold off
```

Lesson 9

Two-D Transient Heat Conduction

This lesson describes 2-D transient heat conduction. In terms of concepts and Matlab coding, there are some minor differences between 1-D and 2-D. However, between 2-D and 3-D, there is no difference in concepts. All the 2-D systems analyzed in this lesson are assumed to be very long perpendicular to the x-y plane with a depth Z.

Nomenclature

nt = total number of time steps
PDE = partial differential equations
q_g^* = a heat-generation-related term, $q_{gen}^{'''} \, \Delta t/(\rho c_v)$, K
r_x = a dimensionless aspect-ratio parameter, $\alpha \Delta t/(\Delta x)^2$
$r_y = \alpha \Delta t/(\Delta y)^2$
Thd = thermodynamics
Z = depth of the control volume, m

---

1. Governing Equation for T(i, j)

The first task for us to accomplish is the derivation of the governing equation for T(i,j).

1-1 General Case

The governing equation for T(i, j) should satisfy the first law of thermodynamics, namely,

$$\Delta U = Q + W,$$

where $\Delta U = mc_v \Delta T$, $W = \Delta x \, \Delta y \, Z q_g^{'''} \Delta t$, and

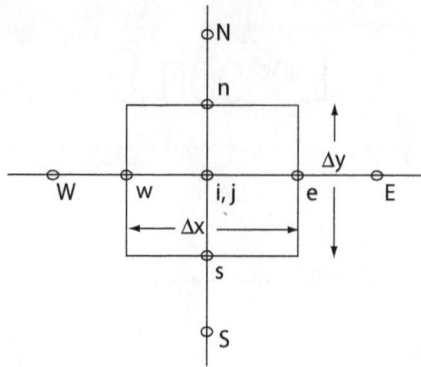

Fig. 9-1   A2-D computational for transient heat conduction

$$Q = (q_w - q_e + q_s - q_n)\Delta t.$$

In reference to Fig. 9-1, the heat flow rate, q_w, for example, can be expressed in terms of the nodal temperatures according to Fourier's law as

$$q_w = k\Delta yZ(T_W - T_{ij})/\Delta x, \text{ etc.}$$

Substituting these expressions into the First Law of Thd eventually yields

$$T_{ij} - T_{ij}^p = r_x (T_W - 2T_{ij} + T_E) + r_y (T_S - 2T_{ij} + T_N) + q_g^*, \tag{1a}$$

or it can be rearranged as

$$-r_x T_W + (1 + 2r_x + 2r_y) T_{ij} - r_x T_E = T_{ij}^p + r_y (T_S + T_N) + q_g^*, \tag{1b}$$

In Eq. (1b), placing T_W and T_E on the left-hand side is to set the equation ready for the numerical line solver.

1-2 Special Cases

(a) Let us be aware that, in deriving Eq. (1b), we have already assumed k = constant. If this assumption is not true, the equation for q_w should be replaced by

$$q_w = k_w \Delta yZ (T_W - T_{ij})/\Delta x$$

instead. Since the lower case w is not a regular node, and k_w must be interpolated by values of k_w and k_{ij}, the equation will become much more complicated.

(b) If $\Delta x = \Delta y$, then $r_x = r_y = r$ is implied. We can rewrite Eq. (1b) as

$$-rT_W + (1 + 4r)\, T_{ij} - rT_E = T_{ij}^p + r\,(T_S + T_N) + q_g^* \tag{2}$$

(c) If $q_g''' = 0$, we further have

$$-rT_W + (1 + 4r)\, T_{ij} - rT_E = T_{ij}^p + r\,(T_S + T_N). \tag{3}$$

(d) If the problem is steady state, implying, $T_{ij} = T_{ij}^p$, Eq. (3) can be further simplified to

$$-T_W + 4T_{ij} - T_E = T_S + T_N. \tag{4}$$

There is a reason we purposely leave T_S and T_N on the right-hand side of the equation. The reason is that both T_S and T_N will be temporarily treated as known quantities during iterations. Such a line-by-line solver generally converges much faster than the point-by-point Gauss-Seidel method. Furthermore, by now, since we have produced several codes for 1-D cases, we should take advantage of their availability.

(e) Two-D fins are merely a special case of 2-D problems.

Two-D fins, shown in Fig. 9-2, are nothing more than systems subject to convective boundary conditions on the top and bottom surfaces. So they are also special cases of 2-D transient problems.

Example 9-1

Consider a 2-D copper column as shown in Fig. 9-3. Initially, $T_W = T_S = T_E = T_N = T_{ij} = 0C$. For copper, the relevant data are given in the Matlab code below. Suddenly, T_W increases to 100C. Write a Matlab code to find $T_{ij}(t)$ for two minutes. Has the copper column reached the steady state yet?

Sol. The governing equation for $T_{ij}(t)$ can be shown to be

$$(1 + 4r)T_{ij} = T_p + rT_W,$$

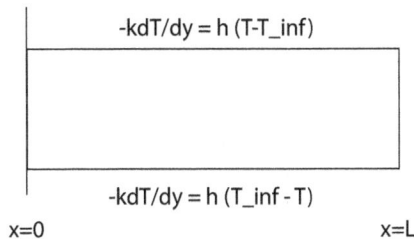

Fig 9-2　A2-D fin is just a special case of 2-D conduction systems

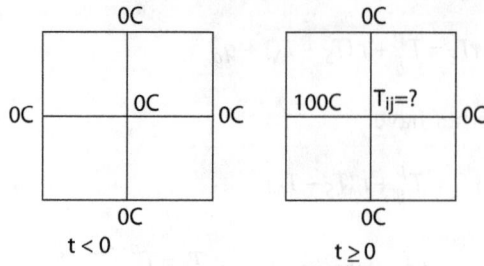

Fig 9-3   A one-node 2-D transient heat conduction system

if the implicit method is used. The Matlab code to compute this one-point problem is given below.

```
clc; clear
Tn = 0; TW=100;
rho=8930; cv=385; k=400; aLf=k/(rho*cv);
dx=0.1; dxs=dx*dx;
Lt=120; nt=20; dt=Lt/nt; ntp=nt+1; % two minutes
r=aLf*dt/dxs;
Tp=0; % initial condition
time(1)=0; T(1)=0;
for it=2:nt+1
time(it)=(it-1)*dt;
T(it)=(Tp+TW*r)/(1+4*r); Tp=T(it);
end
plot(time, T); xlabel('time'); ylabel('T')
hold off
T(ntp) % = 24.82
% check the energy balance in the entire period of heating time
m = rho*dxs;
dU= m*cv*(T(ntp)-T(1)) % = 8.5327e5
sum=0;
for it=2:ntp
sum = sum + k*((100-T(it))-3*(T(it)-0))*dt;
end
Q = sum % = 8.5327e5
```

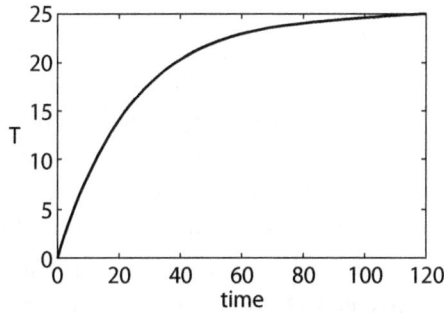

Fig. 9-4   T at the central node as a function of time

As seen in Fig. 9-4, at time = 2 minutes, T has reached 24.82C, almost equal to 25C, which should be the steady state temperature, (100 + 0 + 0 + 0)/4, according to Eq. (4). The energy balance over the entire 2 minutes has also been checked to be true.

Understanding this one-point problem should help to prepare us for understanding a true transient 2-D problem.

2. A Standard Matlab Code for Readers to Modify

Equation (1b) can be readily solved by modifying the 1-D transient heat conduction code. A standard code is presented in the Appendix. It solves a 2-D transient heat conduction problem subject to simple initial condition and boundary condition.

It is recommended that readers understand the 1-D version first. After understanding it, they can comfortably see that the production of the 2-D version is simply adding an iteration loop to improve northern and southern nodal temperature values.

Why do we need to iterate? Because when we first are computing nodal T's at the line j = 2, all values of T_N are garbage. After we have swept computations from j = 2 to j = ny, garbage input produces garbage output. Iterations are required to gradually improve the correctness of these values.

In general, only three major modifications of the standard code are needed to produce our own problem-dependent 2-D codes: (1) give data related to the system and its ambient conditions, (2) establish the numerical grid in time and space, and (3) prescribe initial and boundary conditions. Usually, a good sequence is to prescribe the initial condition first, and then the boundary condition.

3. Speculation on Steel Melting in Concrete Columns During 9/11

A catastrophe happened to the twin World Trade Center towers on 9/11/2001. Two airplanes crashed into the buildings, and the collisions led to the collapse of these two towers. The photo on the right shows the scene of mightily standing towers before the catastrophe. Afterwards, people tried to figure out why WTC towers collapsed. One popular speculation was that the steel columns inside the concrete column melted.

According to the speculation, the jet fuel fires played a very important role in the collapse of the WTC. In this lesson, we will run our Matlab code to roughly estimate the temperature of a typical steel column and to see if the speculation is valid.

The heat conduction problem in the 9/11 tragedy obviously is 3-D. In the north tower, the fire and the explosion started from 93rd floor where the airplane hit the building. If the steel column melted there, we might think that the melting was possible. If it was, let us see if the downward conduction in the steel column would be so rapid that the portion of the steel column on, say, 92nd floor would melt within an hour. Let us assume that there are four steel columns inside a concrete column, as shown in Fig. 9-5, consisting of both a cross-sectional view and a longitudinal view.

Two natural questions arise. (1) How long did it take for the steel column on the 93rd floor to be heated up to its melting point due to cross-sectional conduction via the concrete? (2) How long did it take for the steel column on the 92nd floor to be heated up to its melting point due to the longitudinal conduction via the steel column itself?

Let us split such a 3-D problem into a 2-D one and a 1-D one.

4. Possible Numerical Answers

Figure 9-6 shows the temperature of the steel column as a function of x, parameterized in time. According to the computation performed by running the code in the appendix, if the steel column was at the melting point on the 93rd floor, after one hour, the

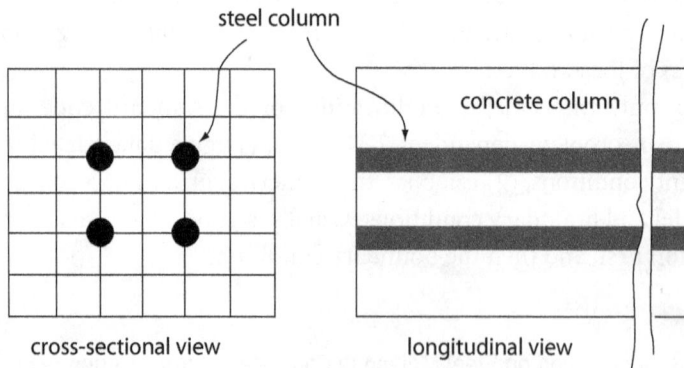

steel column

concrete column

cross-sectional view   longitudinal view

Fig 9-5   A2-D heat conduction system simulating a concrete column embedded with 4 steel columns

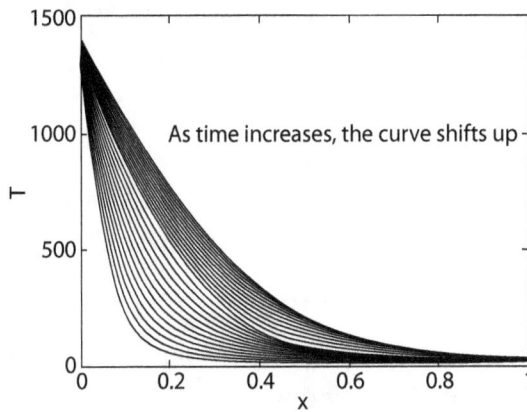

Fig. 9-6   T in the steel column as a function of x parameterized in t.

temperature of the steel column on the 92nd floor would have become somewhat less than 100C, much below 700C, at which the strength of the steel begins to weaken.

See Problem 9-1.

The solution of the 2-D transient heat conduction is shown below. After one hour, the temperatures at the steel columns raised from 20C to only approximately 33.4C.

Based on these two computational analyses, we think that it was unlikely that jet fuel fires played any important roles in the weakening or melting of steel columns.

T_concrete = 1.0e+003 *

		1.0000	1.0000	1.0000	1.0000	1.0000
1.0000	0.2234	0.1338	0.1284	0.1338	0.2234	1.0000
1.0000	0.1338	0.0334	0.0273	0.0334	0.1338	1.0000
1.0000	0.1284	0.0273	0.0212	0.0273	0.1284	1.0000
1.0000	0.1338	0.0334	0.0273	0.0334	0.1338	1.0000
1.0000	0.2234	0.1338	0.1284	0.1338	0.2234	1.0000
		1.0000	1.0000	1.0000	1.0000	1.0000

See Problems 9-2 to 9-5.

5. Use a Two-D Code to Solve a One-D Transient Heat Conduction Problem

In the 2-D code, if we let ny = 2, namely, having only one interior horizontal grid line as shown in Fig. 9-7, and specify the boundary conditions at the bottom and at the top

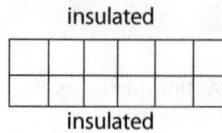

insulated

insulated

Fig. 9-7   A2-D system can be treated as a 1-D system if both bottom and top faces are kept insulated.

to be insulated, namely, $\partial T/\partial y = 0$, the 2-D code is immediately converted to one that is capable of solving 1-D problems, as shown.

In fact, if we desire to solve a steady state problem using this transient code, we can simply let Δt be a large number, and let nt be 2.

6. Exact Solutions for Validation of Codes

To many customers, when they look at the car that they are considering purchasing, the first question arises in their minds (at least in the author's mind) is, does it break down often? Or are there bugs in the design of the machinery of this car?

For users of a computer code they consider running, similarly, the first concern they may have is, is this code bug free? In addition to checking the global and total energy balance of the system, we have another way to convince users that our written codes are likely bug-free.

It is known [1] that all equations, regardless of their complexities and the fact that they are algebraic or differential, can be slightly modified into those whose exact solutions exist and are known to us.

For example, consider the population of fish in a fish pond, x, governed by

$$x^4 - e^x + \sin(x) - 1 = 0. \tag{5}$$

We are interested in finding the solution of x. Equation (5) can be modified into

$$x^4 - e^x + \sin(x) - 1 = c_1, \tag{6}$$

where $c_1 = 15 - e^2 + \sin(2)$ is obtained by substituting $x = 2$ into Eq. (5).

For the real problem of fish population, we do not know if the (or an) exact solution exists. Even if it exists, we do not know what it is.

For the fictitious problem, Eq. (6), however, we do know with certainty that the exact solution, x=2, exists and is even given by us ourselves.

Wisely, when planning to write a code solving a daily life problem, or Eq. (5), we can consider writing a code solving the fabricated problem first, like Eq. (6). When our code is successful in yielding x=2, we have gained confidence in the correctness of our code. Afterwards, all we need to do is to proceed to set c1 to zero, and solve the real problem by running the code.

If the real problem is governed by, for example, two coupled partial differential equations (PDEs) whose unknowns are thermal stress, $S(t, x, y)$, and temperature, $T(t, x, y)$, we can first introduce our chosen exact solution, for example,

$$S(t, x, y) = t^2 + x^2 y^2 \text{ and } T(t, x, y) = t^2 - x^2 y^2. \tag{7}$$

Then we proceed to substitute Eq. (7) into the governing PDEs to obtain two source terms $s_1(t, x, y)$ and $s_2(t, x, y)$.

See Problem 9-6.

When a solution diverges upon running the Matlab code, there are at least three possible sources of errors:

(a) improper modeling,
(b) solutions do not exist,
(c) erroneous programming.

This exact-solution validation procedure can help us to eliminate (c) as a possible error, and focus on errors (a) and (b).

7. Advanced Heat Conduction Problems

Three advanced heat conduction problems will be briefly mentioned here.

7-1 Moving Interface (or called Stefan problem)

Consider a 1-D system consisting of water freezing in a pond, as shown in Fig. 9-8. When studying the equations below, imagine that the system rotates 90 degrees counterclockwise.

If the interface is stationary, then

$$-k_{ice} \left(\frac{dT}{dx}\right)_{ice} = -k_w \left(\frac{dT}{dx}\right)_w \tag{8a}$$

If the interface is moving, then

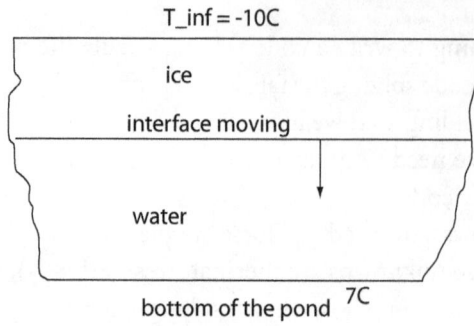

Fig. 9-8   The system schematic of an interface-moving problem

$$-k_{ice}\left(\frac{dT}{dx}\right)_{ice} + \frac{\rho H \Delta x}{\Delta t} = -k_w\left(\frac{dT}{dx}\right)_w \tag{8b}$$

where H is the heat of freezing for ice (exothermic, hence freezing can be considered as heat generation), and $\Delta x/\Delta t$ is the speed of interface moving. Here we assume that the interface moves for Δx exactly within Δt. In reality, it may not.

7-2. Irregular Geometries

Prior to 1970s, finite element methods were superior to finite difference methods in the area of handling irregular geometries. After 1970s, body-fitted coordinates [2] were developed, constituting a boost in preference for the latter, as the latter can be more easily understood with much less math required. The formulation of body-fitted coordinates can be outlined as follows:

(a) Transform (x, y) to (ξ, η). This step is entirely unrelated to heat transfer. For example, see Fig. 9-9. Every node on the x-y plane must correspond to a node on the plane.

(b) Transform PDEs in x and y coordinates into PDEs in ξ and η coordinates

(c) Use C grid, H grid, or O grid.

(d) Solve the transformed PDEs on the ξ - η plane.

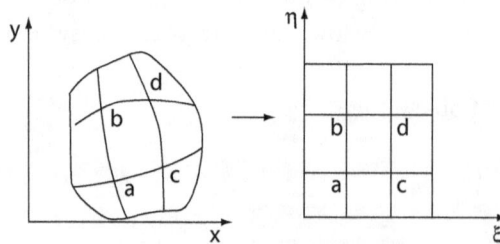

Fig. 9-9   Transformation from an irregular grid to a regualr grid

7-3. Non-Fourier Law (Hyperbolic-Type Heat Conduction Equation)

The classical Fourier law that we have learned so far from Lesson 1 to this lesson,

$$q'' = -k \frac{\partial T}{\partial x}$$

is generally valid for the majority of practical heat conduction situations. In some circumstances, such as during extremely short periods of time under extremely high-flux laser heating, this classical law may break down. During the past few years, considerable attention has been concerned with using the non-Fourier heat conduction law [4, 5], given as:

$$\tau \frac{\partial q''}{\partial t} + q'' = -k \frac{\partial T}{\partial x} \; , \tag{9}$$

where τ is the relaxation time. Differentiating Eq. (9) with respect to x yields:

$$\tau \frac{\partial}{\partial t} \left(\frac{\partial q''}{\partial x} \right) + \frac{\partial q''}{\partial x} = -k \frac{\partial^2 T}{\partial x^2} \; . \tag{9a}$$

On the other hand, energy conservation over an infinitesimal control volume dx yields:

$$\rho c_\nu \frac{\partial T}{\partial t} + \frac{\partial q''}{\partial x} = 0, \tag{10}$$

or $\quad \dfrac{\partial q''}{\partial x} = -\rho c_\nu \dfrac{\partial T}{\partial t} \; . \tag{10a}$

Substituting Eq. (10a) into Eq. (9a) leads to

$$\tau \frac{\partial^2 T}{\partial t^2} + \frac{\partial T}{\partial t} = \alpha \frac{\partial^2 T}{\partial x^2} \; , \tag{11}$$

which is seen to be of hyperbolic type. If the first term in the left-hand side of Eq. (11) is small and can be neglected, Eq. (11) becomes the traditional 1-D heat conduction equation.

This juncture marks the end of our heat conduction studies in this textbook. In the next lesson, we will enter the subject of heat convection.

8. Summary

This lesson can be summarized as:

(a) introduction of a standard Matlab code solving a simple 2-D transient problem,
(b) examination of the thermal aspect of steel columns in the concrete columns during the 9/11 catastrophes,
(c) how to compute 1-D problems or steady-state problems using the transient 2-D code,
(d) a debugging procedure by introducing a fictitious problem whose exact solution exists and is known to us,
(e) briefly mentioning three advanced heat conduction problems

9. References

1. http://911research.wtc7.net/mirrors/guardian2/wtc/how-hot.htm
2. Tien-Mo Shih, A procedure to debug computer programs, *Int. J. Numer. Methods in Eng.*,vol. 21, pp. 1027–1037, 1985
3. J. F. Thompson and Z. U. A. Warsi, Boundary-fitted coordinate systems for numerical solution of partial differential equations—a review, *J. Compu. Physics*, vol. 47, pp. 1–108, 1982.
4. J. Y. Lin, The non-Fourier effect on the fin performance under periodic thermal conditions, *Applied Mathematical Modelling*, vol. 22, pp. 629–640, 1998.
5. P. J. Antaki, Importance of non-Fourier heat conduction in solid-phase reactions, *Combustion and Flame*, vol. 112, pp. 329–341, 1998.

10. Exercise Problems

9-1 Let us revisit the 2-point problem in lesson 6 (2-D steady state). Initially, the brick column is at 500K throughout the system. Suddenly, the boundary temperatures at the bottom decrease to 400K and 300K. Find T1(t) and T2(t). How long does it take for T1 and T2 to reach 460K and 440K? Use the implicit method and the Cramer's rule (instead of Matlab command, q = a\b').

9-2 Use the standard code given in the appendix, and repeat Problem 9-1. Compare the two results.

9-3 Add a small post-processing procedure to check the energy balance globally and totally over the entire 2-D concrete system. Is $\Delta U = Q$, in reference to Problem 9-1?

9-4 In the 1-D transient code for the steel column, the cross-sectional area, Ac, is not included in the computation. Comment on such an exclusion of Ac consideration.

9-5 Consider a 2-D copper column initially at T = 200C and Lx = Ly = 0.01m. Suddenly, a fluid flow at = 25C is blown over the column. The heat transfer coefficient is h = 100 . Take nx=ny=10. Does the system behave like a lumped-capacitance system?

11. Appendix

A-1 A Standard Transient 2-D Code for Readers to Modify

```
clc; clear % a standard code for readers to modify
Lx=1; Ly=2;
nx=10; ny=10; nxp=nx+1; nyp=ny+1; dx=Lx/nx; dy=Ly/ny;
dxs=dx*dx; dys=dy*dy;
rx=.3; ry=.2;
for i=1:nxp
for j=1:nyp
x(i,j)=(i-1)*dx; y(i,j)=(j-1)*dy;
Tp(i,j)=0; T(i,j)=Tp(i,j);
a(i,j)=0; b(i)=0;
end
end
% specify BC
T(1,ny/2+1)=100;
for it=2:2
for iter=1:10
for j=2:ny
 a(1,1)=1; b(1)=T(1,j);
 a(nxp,nxp)=1; b(nxp)=T(nxp,j);
for i=2:nx
 a(i,i-1)=-rx; a(i,i)=1+2*rx+2*ry; a(i,i+1)=-rx;
b(i)=Tp(i,j)+ry*(T(i,j-1)+T(i,j+1)); % treating TS and TN as knowns
end
q=a\b'; T(:,j)=q; % update for iteration
end
end
```

```
Tp=T; % update for time marching
end
T_real=flipud(T')
x_real=flipud(x'); y_real=flipud(y');
mesh(x,y,T)
```

```
%>>>>>>>>>>
```

A-2. Investigation of Transient 1-D Heat Conduction for a Steel Column

```
clc; clear
%>>>>>>>>>> input data of 1-D steel column for the 9/11 tragedy
T_melt=1400;
k=60; rho=8000; cv=477; aLf=k/(rho*cv);
L=1; nx=50; dx=L/nx; dxs=dx*dx; nxp=nx+1;
Lt=3600; nt=20; dt=Lt/nt;
r=aLf*dt/dxs;
x=linspace(0,L,nxp);
%>>>>>>>>>> initialize a and b, and prescribe the B. C.'s
a=zeros(nxp,nxp); b=zeros(1,nxp);
a(1,1)=1; b(1)=T_melt;
a(nxp,nx)=-1; a(nxp,nxp)=1;
%
%>>>>>>>>>> initial condition
for i=1:nxp
Tp(i)=20;
end
time(1)=0;
%
%>>>>>>>>>> interior-T gov. eq.
for i = 2:nx
a(i,i-1)= -r; a(i,i)= 2*r + 1; a(i,i+1)= -r;
end
for it=2:nt+1
time(it)=(it-1)*dt;
for i=2:nx; b(i)= Tp(i);end
T=a\b'; Tp=T;
T1(it)=T(1); Tnxp(it)=T(nxp); % = store up values of T1 and Tnxp
plot(x,T); hold on; xlabel('x'); ylabel('T')
```

text(0.001, 130, 'As time increases, the curve shifts up')
end
%plot(time, T1)
hold off
%>>>>>>>>>> check the total energy balance
% sum1=0; sum2=0;
% for it=2:nt+1
% sum1 =sum1+(qs-h*(Tnxp(it)-Tinf))*dt;
% end
% for i=2:nx
% sum2=sum2 + dx*rho*cv*(T(i)-(-2));
% end
% q_total = sum1 % = 1.9901e6 J/m^2
% dU=sum2 % = 1.9901e6 J/m^2

%>>>>>>>>>>

A-3. Investigation of Transient 2-D Heat Conduction for the Concrete Column

```
clc; clear % steel melting, 2-D, 9/11
k=0.8; rho=2100; cv=1100; aLf=k/(rho*cv);
Lx=0.6; Ly=0.6;
nx=6; ny=6; nxp=nx+1; nyp=ny+1; dx=Lx/nx; dy=Ly/ny;
dxs=dx*dx; dys=dy*dy;
Lt=3600; nt=20; dt=Lt/nt;
rx=aLf*dt/dxs; ry=rx;
for i=1:nxp
for j=1:nyp
x(i,j)=(i-1)*dx; y(i,j)=(j-1)*dy;
Tp(i,j)=20; T(i,j)=Tp(i,j);
a(i,j)=0; b(i)=0;
end
end
% specify BC
for j=2:ny; T(1,j)=1000; T(nxp,j)=1000; end
for i=2:nx; T(i,1)=1000; T(i,nyp)=1000; end
for it=2:nt+1
for iter=1:10
for j=2:ny
```

```matlab
a(1,1)=1; b(1)=T(1,j); a(nxp,nxp)=1; b(nxp)=T(nxp,j);
for i=2:nx
 a(i,i-1)=-rx; a(i,i)=1+2*rx+2*ry; a(i,i+1)=-ry;
b(i)=Tp(i,j)+ry*(T(i,j-1)+T(i,j+1)); % treating TS and TN as knowns
end
q=a\b'; T(:,j)=q; % update for iteration
end
end
Tp=T; % update for time marching
end
T_real=flipud(T')
x_real=flipud(x'); y_real=flipud(y');
mesh(x,y,T)
```

Lesson 10

Forced-Convection External Flows (I)

This lesson marks the first time when we will deviate from topics of heat conduction inside solids, and consider heat transfer in moving fluids, known as heat convection. We will study the hydrodynamic aspect of heat convection first. To stimulate our interest, however, we will take a quick look at a daily-life heat-convection problem in the very beginning.

Nomenclature

C_f = dimensionless shear stress
$\tilde{h}$ = enthalpy, J/kg
$\dot{m}$ = mass flow rate, kg/s
$\dot{M}$ = momentum rate, N
pe = Peclet number, $u_\infty \Delta y/\alpha$
Pe = Peclet number, $u_\infty L/\alpha$ (upper case based on L)
u = x-direction flow velocity, m/s
v = y-direction flow velocity, m/s
Z = the depth in the z direction, m
δ = boundary-layer thickness, m
ξ = dimensionless y, defined as y/ δ
τ = shear stress, Pascal or N/m^2

1. Soup-Blowing Problem

Consider a bowl of soup at T_1 = 80C sitting on the dinner table, as shown in Fig. 10-1. If the air above the soup is perfectly quiescent at T_3 = 20C, then we can assume that the system is approximately a 1-D pure conduction problem, and quickly obtain T_2 to

be 50C. Though inaccurate, this value can serve for comparison later.

A question naturally arises. When blowing cold air over the bowl of soup, why are we able to cool down the soup faster? Let us estimate T_2 under such a situation. Explain our finding to our family members, and imagine that they will all look at us in awe and with admiration.

(a) Step 1: understand the problem statement and start brainstorming

Unlike in the subject of pure heat conduction, there are now additional unknown variables u and v. What forces drive u and v? Obviously, when we blow the soup, high air pressure is created near our lips. Thus, air pressure, p, is involved. If p and T are unknown variables, the density, ρ, will also sneakily enter the problem.

As every state in the U.S. needs a governor, every unknown variable needs a governing equation, too. The temperature, T, is governed by energy conservation; u and v are governed by Newton's law of motion. What about p and ρ? Who governs them? Perhaps we are temporarily overwhelmed by such complexity of the problem. If so, let us regress and attempt to solve a simplified problem first.

(b) Step 2: simplify the problem (also known as modeling)

At this moment, we desire only to make a quick estimate of T_2, and to confirm if T_2 will indeed become lower than 50C. Perhaps we can make several approximations, so that we are able to avoid math and Matlab coding, and see just the qualitative trend.

With this purpose in mind, let us assume:

(i) $u = u_\infty$ = constant,

(ii) $v \ll u$,

(iii) ρ = constant,

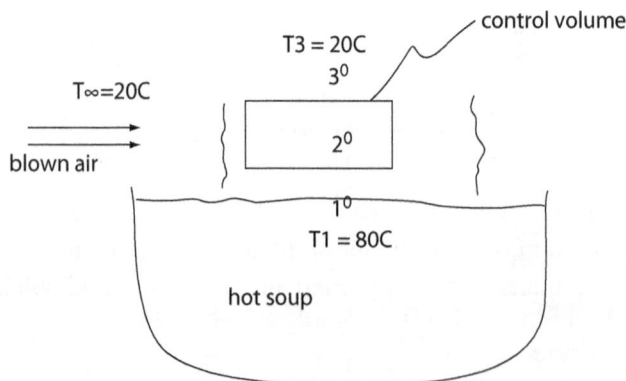

Fig 10-1   System schematic of a bowl of soup

(iv) heat conduction in the x direction << heat conduction in the y direction.
(v) there is only one interior node.
(vi) steady state
(vii) radiation loss from the soup is neglected.
(viii) evaporation effect is neglected.
(ix) the problem is a 2-D one in Cartesian coordinates.

Among these assumptions, (iii) and (iv) are somewhat valid even for real-life situations.

(c) Step 3: derive governing equation for T_2 (still considered as modeling)
An unknown needs a governing equation. A governing equation obeys a law. For the unknown T_2, the law is the first law of thermodynamics. There are two versions of this law: one for closed systems, and the other for open systems—"closed" or "open" in reference to mass flows. Since there are flows into and out from the control volume surrounding T_2, we must adopt the open-system version, namely,

$$\dot{m}_{in}\tilde{h}_{in} - \dot{m}_{out}\tilde{h}_{out} + \dot{Q}_{in} = 0, \tag{1}$$

where

$$\dot{Q}_{in} = \Delta x \, Zk \, (\frac{T_1 - T_2}{\Delta y} - \frac{T_2 - T_3}{\Delta y}), \text{ Z is the depth of the control volume, and}$$

$$\dot{m}_{in}\tilde{h}_{in} - \dot{m}_{out}\tilde{h}_{out} = \rho \, u_\infty \Delta y \, Z \, c_p \, (T_\infty - T_2).$$

Substituting these terms into the first law and assuming $\Delta x = \Delta y$ and $Z = 1$ for simplicity (accuracy is far away from us in this problem anyway) yields

$$T_2 = \frac{100 + 20 \, pe}{2 + pe}, \tag{2}$$

where pe is known as Peclet number, defined as $pe = \frac{u_\infty \Delta y}{\alpha}$. Again, let us reserve Pe for $\frac{u_\infty L}{\alpha}$.

(d) Step 4: carry out the solution procedure to find T_2
Since there is just one unknown in this problem, we do not need to rely on any numerical scheme or Matlab codes. Had there been 1000 nodes, we would need to.

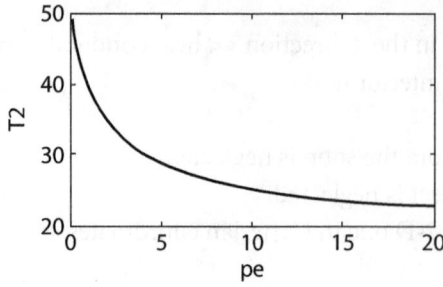

Fig. 10-2   T$_2$ versus pe for the soup problem

(e) Step 5: do the post-processing

The numerical result of T_2 versus *pe* can be plotted in Fig. 10-2. It is seen in the figure that the value of T_2 should lie between 20C (as we blow the soup very hard, and *pe* increases to infinity) and 50C (as the air becomes quiescent).

What are other quantities or trends that can possibly interest us?

Yes, the heat transfer coefficient, h. Are we able to find h? Without exaggeration, this question may be the most important one to ask us ourselves and to answer in the entire subject of heat transfer. Let us delay the analysis till Lesson 12. Now let us turn our attention to some groundwork regarding hydrodynamics first.

2. Boundary-Layer Flows

Consider an isothermal fluid flow moving over a flat plate. There will be a thin layer established over the plate as shown in Fig. 10-3, and is called boundary layer. The flow inside this region is called a boundary-layer flow. More should be said about this flow:

(a) Inside the layer, u, v, and p are functions of x and y. Outside the layer, they are mostly constants.

(b) u >> v and $\partial/\partial y$ >> $\partial/\partial x$ except in the vicinities of the leading edge and the trailing edge.

Example 10-1

At x_{ij} = 5 cm and y_{ij} = 0.1cm with Re_x = 10000, are the two characteristics shown above true for u(x, y)? Take $\Delta x = \Delta y$ = 0.01cm, and the data, which are shown below and are positioned according to the flow situation:

$$u_n^* = 0.68132$$
$$u_w^* = 0.63029 \quad u_{ij}^* = 0.62977 \quad u_e^* = 0.62922, \quad \text{also} \quad u_{ij}^* = 206.65 v_{ij}^*$$
$$u_s^* = 0.57477$$

Sol: From the data above, we find

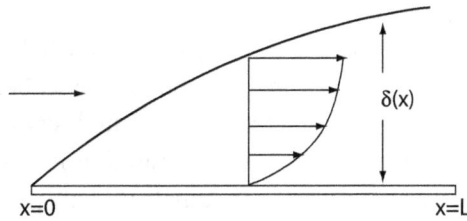

Fig. 10-3   Boundary-layer flow with the thickness exaggerated

$\partial u^*/\partial x = (-0.63029 + 0.62922) / 0.02 = -0.107.$
$\partial u^*/\partial y = (-0.57477 + 0.68132) / 0.02 = 5.3275.$

See Problem 10-6.

(c) The boundary-layer edge is just an imaginary line, artificially drawn to divide the flow into a boundary-layer flow region (in which the viscous effect prevails) and a potential flow region.

See Problem 10-5.

(d) In potential flows, if u_∞ is uniform, then p is uniform due to Bernoulli's equation, suggesting p3 = p4, shown in Fig. 10-4. Based on statement (b), we can safely assume that p1 ≈ p3 and p2 ≈ p4. Therefore, we conclude p1 ≈ p2, suggesting that dp/dx inside the boundary layer is very small in comparison with momentum rate changes and shear stresses, and can be neglected.

See Problems 10-8 and 10-9.

(e) The thickness of the boundary layer grows as x increases. It can be shown later that it is proportional to $x^{1/2}$. It is actually very thin. Most of the figures showing $\delta(x)$ are grossly exaggerated for the sake of clarity.

(f) The transverse velocity, v, is positive everywhere.

See Problem 10-1.

Among the three field variables, u, v, and p, the streamwise velocity u is most important, and therefore will be considered first.

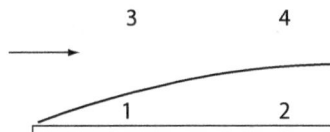

Fig. 10-4   dp/dx is ~ zero

In finding T_2 in the soup problem, we only need a single linear equation, Eq. (1). Finding a single value of u, however, may not be as simple, because the momentum rates involve u^2 and uv, which are nonlinear. Besides, nodes of u and v may have to be staggered to facilitate the mass-conservation analysis, thus increasing the complexities.

Fortunately, a simple way to find an approximate velocity profile is possible.

3. A Cubic Velocity Profile

Let us seek a polynomial profile for u/u_∞ [1]. A quadratic profile may be too crude; a fourth-degree profile may be too algebraically involved.

3-1 Determine the coefficients

Example 10-2

Let us introduce $u^* = \dfrac{u}{u_\infty} = c_0 + c_1\xi + c_2\xi^2 + c_3\xi^3$ as an assumed dimensionless velocity profile, where $\xi = y/\delta$.

Find: values of these four constants.

Sol: Figure 10-3 shows a boundary layer over a flat plate. The thickness, δ, is blown up out of proportion for the sake of clarity. In real life, it is only about 1% of x, depending on the magnitude of Re.

Inside the boundary layer, both u and v are functions of x and y. Outside the boundary layer, the flow is non-viscous, and can be considered as a potential flow. In addition, if the object is a flat plate, u_∞ is uniform.

Relatively, it is a difficult task to find u(x,y) and v(x,y) inside the boundary layer, primarily because the momentum rates, involving u^*u or u^*v, are nonlinear terms. Fortunately, it is possible to find a good approximation of u^* simply by imposing four boundary conditions at y=0 and y=δ. Thus, judging from the sketched velocity profile shown in Fig. 10-3, we assert:

At ξ=0, $u^* = 0 \rightarrow c_0 = 0$.
At ξ=1, $u^* = 1 \rightarrow 1 = c_1 + c_2 + c_3$
At ξ=1, the shear stress is zero, or $du^*/d\xi = 0 \rightarrow 0 = c_1 + 2c_2 + 3c_3$.

The last condition can be acquired by our observation that, near the wall, the u profile looks linear in y. Accepting this observation as a reasonable approximation, we have

At $\xi = 0$, $\dfrac{d^2u}{dy^2} = 0 \rightarrow c_2 = 0.$

Solving for c1 and c3, we finally obtain c1 = 1.5 and c3 = - 0.5. Or,

$$\frac{u}{u_\infty} = 1.5\,\xi - 0.5\,\xi^3, \tag{3}$$

which is a good approximation of the dimensionless streamwise velocity profile inside the boundary layer.

See Problem 10-7.

3-2 Application of Eq. (3)

Establishing this cubic-polynomial velocity profile is by no means merely an ordinary analytical exercise. It actually leads to a few important applications described below:

(a) First, Eq. (3) helps us to find

$$\frac{\delta}{x} = 4.6407\,Re_x^{-1/2} \tag{4}$$

See the appendix for details and Problem 10-2 for exercises.

(b) Whether we like it or not, life is full of comparisons. We are constantly being compared, and are constantly comparing other people, and comparing merits or de-merits of our daily-life decisions. Using Eqs. (3, 4) and the physical data of the ambient air and the system:

$u_\infty = 10$ m/s, $v = 1.6e\text{-}5$ m^2/sec, $\rho = 1$ kg/m^3, x = 0.064 m, and Re_x = 40000,

in reference to Fig. 10-5, we can find momentum rates, $\dot{M}_1$, $\dot{M}_2$, $\dot{M}_3$, $\dot{M}_4$, and the shear force, F5. We can further compare their magnitudes against one another. Which term's magnitude is the largest?

Take the depth in z direction as Z=1 m. For example,

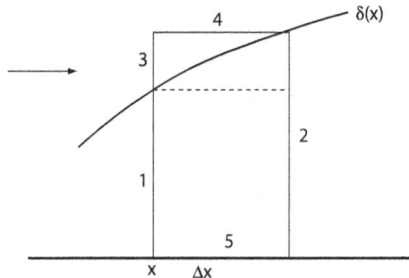

Fig 10-5   **Mass and momentum balances over a slice of control volume. Five components are considered.**

$$\dot{M}_1 = \rho Z \int_0^{\delta 1} u^2 \, dy = \int_0^{\delta 1} u^2 \, dy, \quad \text{and} \quad F5 = \Delta x \, Z \, \mu \left(\frac{\partial u}{\partial y}\right)_{y=0} = \Delta x \, \nu \left(\frac{\partial u}{\partial y}\right)_{y=0}$$

Therefore, all five terms are in the unit of m^3/s^2, enabling us to make comparisons of their magnitudes.

Be very careful in computing $\Delta\delta = \delta2 - \delta1$. The answer is that these five terms ($\dot{M}_1$, $\dot{M}_2$, $\dot{M}_3$, $\dot{M}_4$, and F5) are exactly balanced. Take $\Delta x = 0.01$m. Note that v4 must be found first before $\dot{M}_4$ is computed. This problem trains the rigorousness of our algebraic and arithmetic skills. The procedure is slightly tedious, and will also be described in the appendix.

See also Problem 10-11.

(c) Equation (3) can help us to find the frictional force on the airfoil of an Airbus airplane.

Since shear force on the airfoil = A_airfoil * τ_w and, $\left(\dfrac{du^*}{d\xi}\right)_{\xi=0} = 1.5$, we can obtain

$$\tau_w = \mu \left(\frac{\partial u}{\partial y}\right)_{y=0} = \left(\frac{\mu u_\infty}{\delta}\right)\left(\frac{du^*}{d\xi}\right)_{\xi=0} = 0.3232 \,(\mu u_\infty/x) Re_x^{1/2} \qquad (5a)$$

See Problems 10-3 and 10-4. Often a dimensionless wall shear stress is preferred. This preference leads to

$$C_f = \frac{\tau_w}{\rho u_\infty^2/2} = 0.6465 \, Re_x^{-0.5} \qquad (5b)$$

These values of τ_w and C_f are local ones. They can be readily integrated over [0, Lx] to yield mean values.

(d) Naturally, Eq. (3) can help us to find values of u(x,y) at certain positions inside the boundary layer.

Example 10-3

What is the value of u(x,y) at x = 0.064m and y = 0.00001m, if u_∞ = 10 m/s and Re_x = 40000?

Sol:

clc; clear
rex=4e4; uinf=10; x=0.064; y=1e-5;
delta=4.6407*x/sqrt(rex);
polynomial = 1.5*(y/delta)-0.5*(y/delta)^3;
u = uinf*polynomial % = 0.101 m/s

See Problem 10-10.

(e) It inspires us to apply the same concept to other types of problems in which the profiles look smooth and similar with respect to time and space.

See, for example, the topic of 1-D transient conduction in this textbook.

4. Summary

In this lesson, we have learned:

(a) an elementary cooling problem of a bowl of soup,
(b) definition of boundary-layer flows,
(c) obtaining a cubic profile for the streamwise velocity,
(d) mass balance and momentum balance over a slice of control volume, based on the assumption of the cubic profile.

5. References

1. http://www.see.ed.ac.uk/~johnc/teaching/fluidmechanics4/2003-04/fluids13/apllications.html

6. Exercise Problems

Only laminar airflows over a flat plate are considered here.

10-1 At the boundary-layer edge, is the y-direction flow velocity entering the boundary-layer region or exiting? (In other words, is v_∞ positive or negative?)
(a) positive, (b) negative, (c) zero, (d) it can be either way, depending on the value of Re_L.

10-2 If u_∞ increases, the boundary-layer thickness, δ, will:

(a) increase, (b) decrease, (c) remain the same, (d) increase if Re > 1e4, and decrease if Re < 1e4.

10-3 The local wall shear stress at a certain x position is proportional to:

(a) u_∞ linearly, (b) $u_\infty^{1/2}$, (c) $u_\infty^{3/2}$, (d) $u_\infty^{-1/2}$.

10-4 The local wall shear stress is proportional to:

(a) x linearly, (b) $x^{1/2}$, (c) $x^{-1/2}$, (d) x^0, meaning, independent of x.

10-5 What does the edge line of the boundary-layer region stand for?

(a) a streamline, (b) an imaginary line where u ≈ u_∞, (c) a line where u_∞ ≈ 0, (d) a line to which the velocity vector is tangential.

10-6 Which comparison is qualitatively true, ignoring their signs:

(a) $\dfrac{\partial v}{\partial x} << \dfrac{\partial u}{\partial y}$, (b) $\dfrac{\partial u}{\partial x} << \dfrac{\partial v}{\partial y}$, (c) $\dfrac{\partial v}{\partial x} \approx \dfrac{\partial u}{\partial y}$, (d) $\dfrac{\partial u}{\partial x} >> \dfrac{\partial v}{\partial y}$.

10-7 Instead of a cubic-polynomial velocity profile, a reasonable quadratic-polynomial counterpart should be

(a) $u^* = \xi^2$, (b) $u^* = 0.5\,\xi + 0.5\,\xi^2$, (c) $u^* = 1.5\xi - 0.5\xi^2$, (d) $u^* = 2\xi - \xi^2$.

10-8 Prairie dogs are very cute. Smart, too. It is known that prairie dogs are able to create ventilation effects inside their burrows without paying a penny of electricity bills and without taking heat convection classes. What they do is to dig a tunnel underground as shown in Fig. 10-6. Around one entrance, they pile up a mound of dirt. When wind blows, there will be a counterclockwise ventilation taking place inside the burrow. The primary principle they follow is

(a) mass conservation,
(b) first law of thermodynamics,
(c) Bernoulli's Equation,
(d) law of natural convection.

10-9 On a family outing, you are driving on the freeway at 60 miles/hr. Your younger sister wants to open the car window. You warn her that if she does,

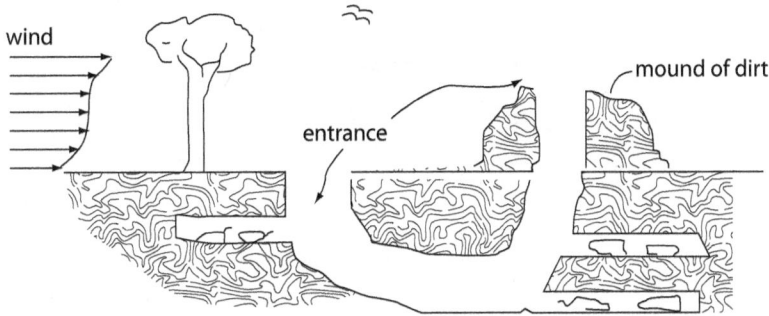

Fig 10-6   Sideview of a prairie dog's underground burrow

she will experience a pressure difference on her eardrums. She is impressed by your fluid mechanics knowledge, listens to your advice, and hence, no argument ensues. The ratio of that pressure difference to the air pressure is approximately:

(a) 10 %, (b) 0.35 %, (c) 0.05 %, (d) 0.01%.

10-10 If the air flow is moving at u_∞ = 10 m/s, with ρ = 1kg/m^3, ν = 1.6e-5 m^2/sec, x=0.064m, and Δy = 3.2e-7m, the streamwise velocity, u, at node 2 in reference to Fig. 10-7 is:

(a) 0.0032 m/s, (b) 0.0064 m/s, (c) 0.014 m/s, (d) 0.028 m/s.

[Problem #11 is not a multiple-choice question]

10-11 Consider air flow moving at u_∞ = 10 m/s, with ρ = 1kg/m^3, ν = 1.6e-5 m^2/sec, x=0.064m, and Δx = 0.006m. To intentionally augment the magnitudes of all the terms, let Z=100m. Find all the mass flow rates as shown in Fig. 10-8. In

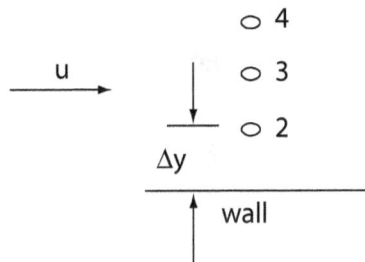

Fig. 10-7   Grid nodes adjacent to the wall

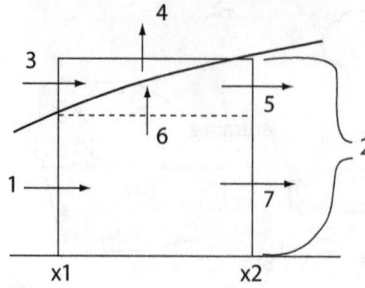

Fig 10-8   Mass balance over the rectangular control volume crossed by mass flows 3, 4, 5, and 6.

particular, please check if the mass is balanced over the small rectangle $(\dot{m}_3 + \dot{m}_6 - \dot{m}_4 - \dot{m}_5) = 0$.

Hint: Some attention must be given to finding $\dot{m}_7$ (or $\dot{m}_5$).

5. Appendix

Let us find mass flow rates first. Keep in mind that we have taken $\rho = 1$ and $Z = 1$.

$$\dot{m}_1 = \int_0^{\delta_1} u\,dy = u_\infty\,\delta_1 \int_0^1 (1.5\xi - 0.5\,\xi^3)\,d\xi = 0.625\,u_\infty\delta_1 \,. \tag{A-1}$$

$\dot{m}_2 = 0.625\,u_\infty\delta_2$, and $\dot{m}_3 = u_\infty\,\Delta\delta$,

where $\Delta\delta = \delta_2 - \delta_1$.

Hence, $\dot{m}_4 = \dot{m}_1 + \dot{m}_3 - \dot{m}_2 = 0.375\,u_\infty\,\Delta\delta$. $\tag{A-2}$

Consequently, Eq. (A-2) leads to

$$v_\infty = (0.375)\,u_\infty\,\Delta\delta/\Delta x. \tag{A-3}$$

Next, let us find those four momentum rates.

$$\dot{M}_1 = \int_0^{\delta_1} u^2\,dy = u_\infty^2\,d_1\int_0^1 (1.5\xi - 0.5\,\xi^3)^2\,d\xi = 0.4857\,u_\infty^2\,\delta_1.$$

Similarly, $\dot{M}_2 = 0.4857\,u_\infty^2\,\delta_2.$ Or $\dot{M}_2 - \dot{M}_1 = 0.4857\,u_\infty^2\,\Delta\delta.$ $\tag{A-4}$

$$\dot{M}_3 = u_\infty^2\,\Delta\delta, \tag{A-5}$$

and

$$\dot{M}_4 = v_\infty u_\infty \Delta x = 0.375 \, u_\infty^2 \Delta \delta. \tag{A-6}$$

Finally, the shear force exerted by the plate on the fluid inside the control volume can be found as

$$\text{wall shear force} = -\Delta x \left(v \frac{\partial u}{\partial y} \right)_{y=0} = -\Delta x \left(\frac{v u_\infty}{\delta_1} \right) \left[\frac{d}{d\xi} \left(\frac{u}{u_\infty} \right) \right]_{\xi=0} = -1.5 \Delta x \left(\frac{v u_\infty}{\delta_1} \right)$$

Since $\dot{M}_2 - \dot{M}_1 + \dot{M}_4 - \dot{M}_3 = $ wall shear force, we obtain

$$\frac{(0.4857 + 0.375 - 1)}{1.5} \frac{\Delta d}{\Delta x} = - \frac{v}{d_1 u_\infty} .$$

Or, in the limit,
$\Delta \delta / \Delta x \to d\delta/dx$ and $\delta_1 \to \delta$, and realizing

$$2\delta \frac{d\delta}{dx} = \frac{d}{dx} (\delta^2),$$

we are able to derive

$$\delta^2 = \left(\frac{3}{0.1393} \right) \left(\frac{vx}{u_\infty} \right) \text{ or }$$

$$\delta = 4.6407 x Re_x^{-1/2}. \tag{A-7}$$

Therefore, for this problem ($x=0.064$m, $u_\infty = 10$ m/s and $Re_x = 40000$)

$$\delta = 0.00587 \, x^{1/2}.$$

In addition, we can also obtain

$$\frac{d\delta}{dx} = \frac{0.00587}{2 x^{\frac{1}{2}}}, \text{ or at } x = 0.064, \Delta \delta = 0.011602 \Delta x, \delta_1 = 0.001485, \text{ and } \frac{\delta_1}{x} = 0.02320.$$

Let us now check if the momentum equation

$$\dot{M}_2 - \dot{M}_1 + \dot{M}_4 - \dot{M}_3 = \text{Shear force?}$$

is exactly balanced:

$$(\text{LHS})/(u_\infty^2 \Delta \delta) = (\dot{M}_2 - \dot{M}_1 + \dot{M}_4 - \dot{M}_3)/(u_\infty^2 \Delta \delta) = 0.4857 + 0.3750 - 1 = -0.13930.$$

$$\text{wall shear force} = -\frac{\Delta x u_\infty^2}{Re_x} \frac{1.5}{0.02320} = -0.13932 \, u_\infty^2 \Delta \delta. \text{ Or,}$$

$$(\text{RHS})/(u_\infty^2 \Delta \delta) = -0.13932. \to \text{LHS} = \text{RHS. Yes, balanced, indeed.}$$

Lesson 11

Forced-Convection External Flows (II)

In this lesson, a few important subjects will be presented. It is beneficial for us to be aware of them before we learn how to solve energy-governing equations to obtain the temperature distribution in the next lesson.

Nomenclature

Ec = Eckert number, defined as $u_\infty^2/(c_p\Delta T)$
Fo = Fourier number, defined as $\alpha\Delta t/L^2$
Gr = Grashof number, defined as $g\beta\Delta TL^3/v^2$
Le = Lewis number, defined as α/D
Nu = Nusselt number, defined as hL/k
Qs = a dimensionless quantity, defined in Eq. (1d)
Re = Reynolds number, defined as $u_\infty L/v$
Sc = Schimdt number, v/D
St = Stanton number, $h/(\rho c_p u_\infty)$
u = x-direction flow velocity, m/s
v = y-direction flow velocity, m/s
ξ = dimensionless x, x/L_x

---

1. Nondimensionalization (abbreviated as Ndm)

Strictly speaking, the use of adjectives is meaningful only when comparisons are made. For example, it is not too meaningful for us to say, "George's house is very big" unless we and our listeners have clear comparisons in mind. Otherwise, our statement will likely be incorrect if George's house is compared with the White House in Washington D.C.

Antidisestablishmentarianism?
It is the longest non-technical word.
28 letters

In lesson 10, we mentioned large convection terms. By the same rationale, we must compare these terms with something in order to make the adjective "large" meaningful. Usually, we compare convection terms with diffusion terms, or in other words, compare momentum terms with stress terms. For example,

$$\rho u^2 \propto \mu \frac{\Delta u}{\Delta y} \ .$$

Upon introduction of a reference velocity u_0 and a reference length L, the expression above becomes

$$Re \, u^{*2} = \frac{\Delta u^*}{\Delta y^*} \ ,$$

where $Re = \frac{u_0 L}{v}, \ u^* = \frac{u}{u_0}, \ and \, y^* = \frac{y}{L}$,

The starred values generally range from 0 to 1. Reynolds number is an indicator of the strength of momentum relative to that of shear stress. Hence, if Re > 1, we may be able to claim that the convection term is large relative to the shear stress term.

See Problem 11-1.

This small illustration logically leads us to the topic of nondimensionalizaiton (Ndm). As we have seen, performing Ndm is clearly an additional algebraic burden for us. We suspect that there must be a few advantages associated with such a task other than mere comparisons of magnitudes of terms. Or else nobody would bother to perform it. These advantages are described below.

At least three advantages of Ndm can be identified.

1-1 Concise Algebraic Expressions

More concise algebraic expressions, with numbers in the neighborhood of 0 and 1, will appear.

Both algebraically and arithmetically, the procedure becomes easier and neater than that dealing with dimensional terms. Let us take a look at a complete solution procedure for a Ndm-ed heat conduction problem.

Example 11-1

Given: 1-D slab heat conduction, steady state, both sides exposed to different air-flows, shown in Fig. 11-1

Find: T(x)

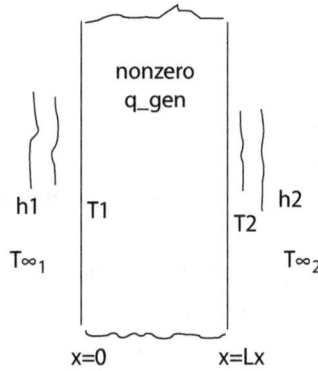

Fig. 11-1   A problem to learn Ndm

Sol:
(i) introduce $\theta = \dfrac{T - T_{\infty 1}}{\Delta T}$ and $\xi = x/L_x$.

where $\Delta T = T_{\infty 2} - T_{\infty 1}$.

Also, $\theta = \theta_1 + d_1 \xi + d_2 \xi^2$.

(ii) There are four unknowns: θ_1, θ_2, d_1, and d_2. Thus, we need to establish four equations. Let us do so below:

definition of θ_2: at $\xi = 1$, $\theta_2 = \theta_1 + d_1 + d_2$. $\hspace{2cm}$ (1a)

boundary condition at x=0: $h_1 (T_{\infty 1} - T_1) = -k \left(\dfrac{\partial T}{\partial x} \right)_{x=0}$,

or $Bi_1 \theta_1 = d_1$ $\hspace{6cm}$ (1b)

boundary condition at x=Lx: $h_2 (T_2 - T_{\infty 2}) = -k \left(\dfrac{\partial T}{\partial x} \right)_{x=Lx}$,
or

$Bi_2 \theta_2 = d_1 + 2d_2$. $\hspace{6cm}$ (1c)

global energy balance: $h_1 (T_{\infty 1} - T_1) + q'''_{gen} Lx = h_2 (T_2 - T_{\infty 2})$, or

$Bi_1 \theta_1 + Bi_2 \theta_2 = Qs + Bi_2$, where $Qs = \dfrac{q'''_{gen} L_x^2}{k \, \Delta T}$ and $Bi = \dfrac{hL}{k}$. $\hspace{1cm}$ (1d)

(iii) We can solve these four equations analytically. It is also our option, however, to take advantage of Matlab.

```
clc; clear
Lx=0.1; k=1; h1=100; h2=200;
Tinf1=0; Tinf2=100; dT=Tinf2-Tinf1;
for igen=1:2 % generate two cases: qgen=0 and qgen=1e5.
qgen=(igen-1)*1e5;
Qs =qgen*Lx*Lx/(k*dT); % qgen non-dimensionalized
Bi1=h1*Lx/k; Bi2=h2*Lx/k;
% The most important equation is the global
% conservation of energy
a=[1 -1 1 1; Bi1 0 -1 0; 0 Bi2 1 2; Bi1 Bi2 0 0];
b=[0 0 Bi2 Qs+Bi2];
% 1 – 2 + d1 + d2 = 0 ................1st row
% Bi1*1 – d1 = 0 ...................... 2nd row
% Bi2*2 +d1 +2*d2 = Bi2 .... ........3rd row
% Bi1*1 + Bi2* 2= Qs + Bi2 ..... 4th row
q=a\b'; th1=q(1); th2=q(2); d1=q(3); d2=q(4);
nx=20; nxp=nx+1; dx=Lx/nx;
for i=1:nxp
 x(i)=(i-1)*dx; ksi=x(i)/Lx;
 th(i)=th1+d1*ksi+d2*ksi*ksi; % find first
 T(i)=Tinf1+th(i)*dT; % recover T from
end
plot(x, T); xlabel('x'); ylabel('T'); hold on
end; text(0.05, 45, 'Qgen = 0'); text(0.04, 195, 'Qgen = 10^5');
```

This method can be viewed as a spectral method, in which basis functions are non-zero everywhere inside the computational domain. By contrast, in finite difference methods and finite element methods, basis functions are non-zero only in a certain grid interval or in a certain element.

See Problems 11-3, 11-4, and 11-5.

Readers can attempt to solve the same problem dimensionally. They will find that the procedure will become more tedious algebraically.

1-2 Identifications of Physical Parameters

In the example above, because of using Ndm, we are able to identify at least two parameters, i.e., Biot number, Bi, and Qs. In other words, in principle, John can use a slab made of copper, with cold air flows blown over both faces. Mary can use a slab

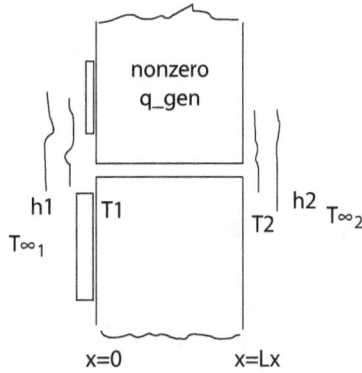

Fig 11-2   A complicated heat-transfer problem

made of carbon, with hot water flows pumped over both sides. In their laboratories, as long as they keep values of Bi1, Bi2, and Qs the same, their values of, say, θ_2, should be measured to be the same.

See Problem 11-2.

1-3 Task Sharing

As a group, we are able to share tasks among engineers, applied mathematicians, and computer software programmers (the most pronounced advantage).

Imagine that a hole is now drilled through the slab and small tiles are placed on the left face of the slab for certain industrial applications shown in Fig. 11-2, such that h1 and h2 are no longer known for such a conjugate (combined heat convection and conduction) problem. It is likely that engineers know how to model the problem, but do not know how to solve the problem computationally. Applied mathematicians know how to solve complicated equations, but may not be familiar with transforming the problem into a set of equations.

Hence, these two groups of specialists need to collaborate and communicate.

An efficient way of collaboration and communication may be the following: an engineer is assigned the physical problem by the VP of his consulting company. She models the problem and produces a set of dimensionless governing equations, which constitute a purely mathematical problem. Then the applied mathematician takes over the mathematical problem, solves it computationally, and takes the computational

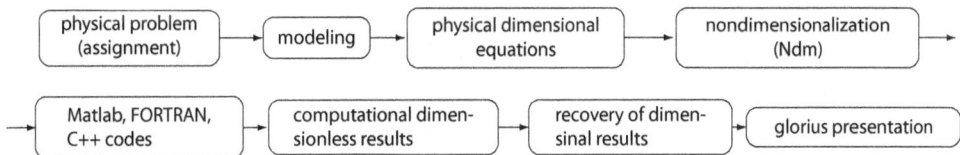

result to the engineer. The engineer subsequently recovers the dimensional quantities from the dimensionless result, and presents them to the VP. The streamline of operation is like the chart shown.

2. Important Dimensionless Parameters in Heat Transfer

It is a good timing for us to list important and popular dimensionless parameters in heat transfer below. A brief explanation is attached, regarding how or when they arise. They include:

$Bi = hL/k_s$, Biot number, convective boundary condition

$C_f = \tau_s/(0.5\rho u_0^2)$, Coefficient of Friction, dimensionless shear stress at the wall

$Ec = u^2/(c_p\Delta T)$, Eckert number, high-speed flows

$Fo = \alpha\Delta t/L^2$, Fourier number, transient heat conduction

$Gr = g\beta\Delta TL^3/v^2$, Grashof number, free convection flows

$Le = \alpha/D$, Lewis number, relative strength between the conduction and the species diffusion

$Nu = hL/k_f$, Nusselt number, dimensionless heat transfer coefficient

$Pe = u_0L/\alpha$, Peclet number, very similar to Re, equal to Pr * Re

$Pr = v/\alpha$, Prandtl number, relative strength between the viscosity and the conduction

$Re = u_\infty L/v$, Reynolds number, relative strength between the momentum and the friction

$Sc = v/D$, Schimdt number, relative strength between the viscosity and the species diffusion

$St = h/(\rho c_p u_\infty)$, Stanton number, a dimensionless heat transfer coefficient

Among these parameters, Ec, Le, Sc, and St will not be used in this textbook. To check if St is indeed dimensionless, for example, we write:

$$St = (W / m^2\text{-K}) / [(kg / m^3) (J / kg\text{-K}) (m / s)] = 1.$$

3. Derivation of Governing Equations

It is often a dilemma for heat-transfer instructors to make whether or not they should include formal derivations of governing equations in their lectures to the classes. If these derivations are covered, the material tends to be dry and full of equations, and students generally do not accept it well. If they are not covered, however, students will not fully understand how terms in the governing equations arise and what they represent. Unfortunately, understanding the governing equations usually is the starting point of an analysis for heat-transfer problems.

In this textbook, we have decided to leave these derivations to the appendix. Both instructors and students can make their own choices if they desire to devote their time to reviewing them.

4. Categorization

Finally, before turning our attention to methods of finding temperature distributions, we will become aware of various common types of heat-convection problems.

In terms of:
- flow pattern,there are (a) laminar (b) turbulent
- solid boundary, (a) external flows, (b) internal flows
- main driving force, (a) forced convection, (b) free convection (or natural convection)
- time dependence, (a) steady state, (b) unsteady state
- compressibility, (a) incompressible flows, (b) compressible flows
- combined phenomena, (a) with conduction, (b) with radiation, (c) with mass transfer, (d) with combustion, (e) with many others
- species, (a) single species, (b) multi-species
- phases, (a) single phase, (b) multi-phases

Example 11-2

What is the conventional definition of an incompressible flow? Is a bonfire a compressible flow?

Sol: In general, in a flow system, if $\Delta p \ll p$, or Ma 0.3, then the flow is considered an incompressible flow [1].

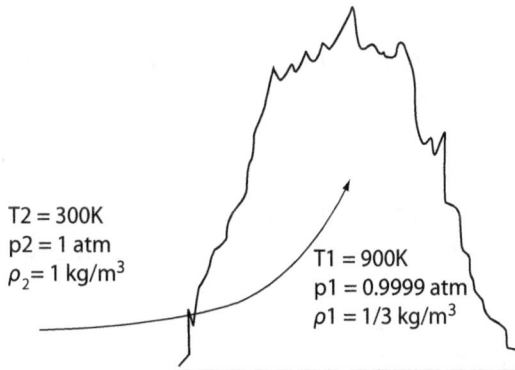

$T2 = 300K$
$p2 = 1$ atm
$\rho_2 = 1$ kg/m^3

$T1 = 900K$
$p1 = 0.9999$ atm
$\rho1 = 1/3$ kg/m^3

Fig. 11-3   A bonfire is an incompressible flow

Consider a bonfire shown in Fig. 11-3. Let us estimate the flow velocity of the entrained air to be 10 m/s. Hence,

$0.5\rho u^2 = 0.5*1*10*10 = 50$ Pa.

According to Bernoulli's equation, we have

$p_2 - p_1 \approx 0.5\rho u^2 = 50$ Pa $<<$1e+5 Pa $\approx p$.

So, a bonfire should be considered an incompressible flow, even though its ρ is a variable.

5. Summary

This lesson presents:

(a) Nondimensionalization (Ndm) and reasons that it is often desired,
(b) important physical parameters commonly encountered in heat convection,
(c) categorization of heat-convection problems,
(d) derivation of governing equations in both the algebraic form and the differential form.

6. Reference

1. John D. Anderson, *Modern Compressible Flows*, McGraw-Hill, 2004.

7. Exercise Problems

11-1 Let us revisit Example 10-1. The values shown in the example are dimensionless quantities, in reference to u_∞.

(a) Recover the dimensional u's and v_{ij}.
(b) Show that the following equation, in its algebraic form, will be satisfied after appropriate numbers are substituted into the following equation. Take $\nu = 1.6$e - $5 m^2/sec$.

$$u \frac{\partial u}{\partial x} + v \frac{\partial u}{\partial y} = \nu \left(\frac{\partial^2 u}{\partial x^2} + \frac{\partial^2 u}{\partial y^2} \right)$$

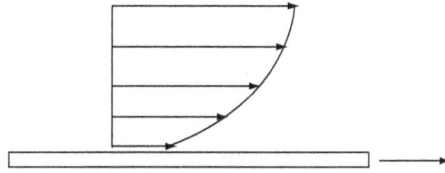

Fig. 11-4   Velocity profile with slip at y = 0

(c) It is also important to possess a general sense of both the magnitude and the sign of individual terms.

(d) Ponder over the situation in which you are asked to find u_{ij}, if u_{ij} had been the ONLY unknown in the algebraic equation above. If you are able to find it, wonderful. If you are unable, justify why you are unable.

11-2. Consider the following scenario: you are working for a heat-transfer consulting firm and are the technical leader of a design team. The CEO of your company would like to delegate you to solve a problem pertaining to a circular one-D fin exposed to the clear sky overnight.

At 6 pm, or t =0, the initial condition is T(0, x) = 100C throughout the fin. The CEO is interested in finding T(t, x) during 12 hours between 6 pm and 6 am the next day. T_b and T_∞ will be maintained at 100C (373K) and 5C (278K), respectively, all the time. Knowing that there will be radiation loss that complicates the problem, the CEO assigns Bob, who knows only math, but no heat transfer, to help you.

Derive the dimensionless governing equation and associated boundary condition (insulated at the fin tip) that you will hand to Bob. Identify pertaining parameters. The equation can be either differential or algebraic. To unify our solutions, let us introduce $\varphi = T/T_\infty$ but not $\theta = (T - T_b)/(T_\infty - T_b)$, since there is also a radiation term.

It is ultimately important that, in the document to Bob, there should not exist a single tiny term that is dimensional; otherwise Bob will be confused/upset, the assignment will be delayed, and your job security will be affected. Note that even t is dimensional. But $t^* = t/t_0$ is dimensionless.

Hint: The final dimensionless equation should contain four parameters, no more, no fewer:

$r = \alpha t_0/L^2$, Bi = hL / k, $A^* = pL/A_c$ (area ratio), and $\beta = (t_0 T_\infty^3 p \in \sigma)/(\rho c_\nu A_c)$.

The time reference, t_0, can be chosen to be 12 hours, for example.

11-3. Revisit the dimensionless one-D slab problem given in Example 11-1.

(a) Run the Matlab code, as is, to find $T(x = 0.5Lx)$.
(b) Let us replace the slab with one having $k = 0.1$ W/m-K. Adjust values of given data such that the new T value remains the same as that in (a).

11-4 Consider a boundary-layer flow system in which the plate is also moving, at a relatively slow speed, u_0, and in the same free stream direction, as shown in Fig. 11-4. Using the concept of Ndm and cubic-polynomial profile, find γ, where

$$u_{mean} = \gamma u_\infty + (1 - \gamma) u_0.$$

11-5 Given: air boundary-layer flow, $u_\infty = 10$ m/s, $Re_x = 40000$, $x = 0.064$m, and $L = 0.16$m. When Bob brings back to you his computer output containing all $u^*(i, j)$ and $v^*(i, j)$, what value do you expect $u^*(i, 2)$ to be if $\Delta y^* = 2e\text{-}6$?

(a) 0.00075, (b) 0.0075, (c) 3.2e-5, (d) 3.2e-4

8. Appendix: Steady-State Governing Equations

Details of deriving the momentum equation in x direction will be presented in this section.

A-1. Derivation of Governing Equation for u

Let us start from Newton's second law of motion, $F = m\,a$,

or $\Sigma F = dM/dt$ — Lagrangian view. $\hspace{3cm}$ (A-1)

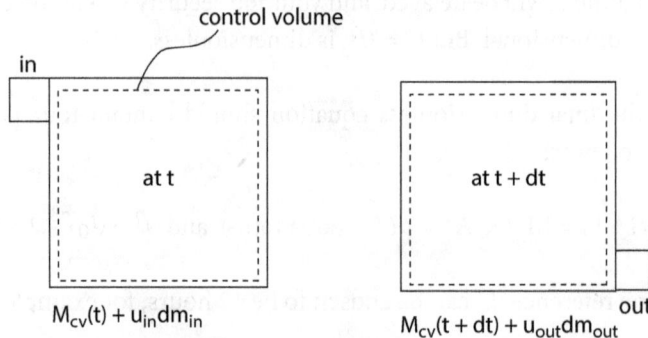

$$M_{cv}(t) + u_{in}dm_{in} \hspace{3cm} M_{cv}(t + dt) + u_{out}dm_{out}$$

Fig. 11-5   Conversion from closed system to open system

In heat convection, it is convenient for us to convert this law in the Lagrangian view to the Eulerian view, primarily because we do not wish to follow the air particles around.

Consequently, according to Fig. 11-5,

$$\left(\frac{dM}{dt}\right) \approx \frac{1}{\Delta t}\left[M_{cv}(t + \Delta t) + u_{out}(\Delta m)_{out} - M_{cv}(t) - u_{in}(\Delta m)_{in}\right].$$

In the limit $\Delta t \to 0$, we obtain

or $\Sigma F = \left(\frac{dM}{dt}\right)_{cv} + \dot{m}_{out}\,u_{out} - \dot{m}_{in}\,u_{in}$. ----- Eulerian view. (A-2)

For steady-state problems, $(dM/dt)_{cv}$ vanishes. If there are no mass flows across the boundary of the control volume, Eq. (A-2) becomes Eq. (A-1).

The mass flow rates entering and leaving the control volume can be derived as

$\dot{m}_w = \rho_w u_w \Delta y Z$, $\dot{m}_s = \rho_s v_s \Delta x Z$.

$\dot{m}_e = \rho_e u_e \Delta y Z$, and $\dot{m}_n = \rho_n v_n \Delta x Z$.

If we assume that Z=1m deep and ρ is constant, we obtain, in reference to Fig. 11-6,

Momentum rate out – Momentum rate in

$= \dot{m}_e u_e + \dot{m}_n u_n - \dot{m}_w u_w - \dot{m}_s u_s$

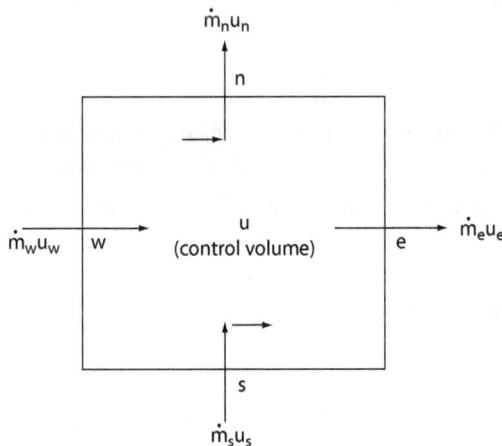

Fig. 11-6   **Momentum in and out**

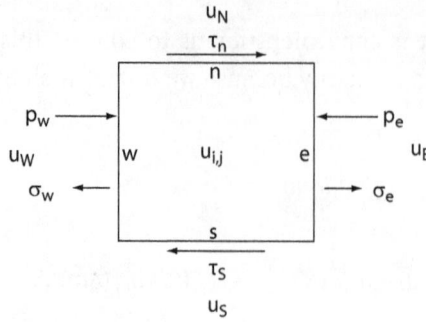

Fig. 11-7 **Force balance over a control volume**

$$= \rho \Delta y (u_e^2 - u_w^2) + \rho \Delta x (v_n u_n - v_s v_s).$$

Also, according to Fig. 11-7,

$$\Sigma F = (p_w - p_e)\Delta y Z + (\sigma_e - \sigma_w)\Delta y Z + (\tau_n - \tau_s)\Delta x Z.$$

But $\sigma_e = \mu \dfrac{(uE - uij)}{\Delta x}$, $\sigma_w = \mu \dfrac{(uij - uw)}{\Delta x}$, (imagine the control volume as a chunk of chewing gum),

$$\tau_n = \mu \frac{(uN - uij)}{\Delta x}, \text{ and } \tau_s = \mu \frac{(uij - us)}{\Delta y}.$$

Finally, realizing that $M_{out} - M_{in} = \Sigma F$, and dividing both sides of the equation by $\rho \Delta x$ Δy, we obtain

$$(u_e^2 - u_w^2)/\Delta x + (v_n u_n - v_s v_s)/\Delta y = \nu(u_w - 2u_{ij} + u_E)/(\Delta x)^2 + \nu(u_s - 2u_{ij} + u_N)/(\Delta y)^2$$

$$-(-p_w + p_e)/(\rho \Delta x). \tag{A-3}$$

If we pay close attention to the equation above, subscripts of the momentum terms and the pressure term are located at w, s, e, and n on the boundary of the control volume, whereas subscripts of normal stresses and shear stresses are located at W, S, E, and N at adjacent grid nodes.

In the limit of diminishing $\Delta \mathbf{x}$ and $\Delta \mathbf{y}$, we obtain

$$\frac{\partial}{\partial x}(u^2) + \frac{\partial}{\partial y}(vu) = \nu\left(\frac{\partial^2 u}{\partial x^2} + \frac{\partial^2 u}{\partial y^2}\right) - \frac{1}{\rho}\left(\frac{\partial p}{\partial x}\right). \tag{A-4a}$$

Using the continuity equation, we can also modify the convective terms into

$$u\frac{\partial u}{\partial x} + v\frac{\partial u}{\partial y} = v\left(\frac{\partial^2 u}{\partial x^2} + \frac{\partial^2 u}{\partial y^2}\right) - \frac{1}{\rho}\left(\frac{\partial p}{\partial x}\right). \tag{A-4b}$$

A dimensionless version of the equation above can be derived as follows:

$$\frac{u_0^2}{L}\left(u^*\frac{\partial u^*}{\partial x^*} + v^*\frac{\partial u^*}{\partial y^*}\right) = \frac{vu_0}{L^2}\left(\frac{\partial^2 u^*}{\partial x^{*2}} + \frac{\partial^2 u^*}{\partial y^{*2}}\right) - \frac{1}{\rho}\frac{\partial p}{\partial x} \quad \text{or}$$

$$\left(u^*\frac{\partial u^*}{\partial x^*} + v^*\frac{\partial u^*}{\partial y^*}\right) = \frac{1}{Re}\left(\frac{\partial^2 u^*}{\partial x^{*2}} + \frac{\partial^2 u^*}{\partial y^{*2}}\right) - \frac{\partial p^*}{\partial x^*}, \tag{A-5a}$$

where $p^* = p/(\rho u_0^2)$. By a similar procedure, we can obtain the momentum equation for v^* as

$$\left(u^*\frac{\partial v^*}{\partial x^*} + v^*\frac{\partial v^*}{\partial y^*}\right) = \frac{1}{Re}\left(\frac{\partial^2 v^*}{\partial x^{*2}} + \frac{\partial^2 v^*}{\partial y^{*2}}\right) - \frac{\partial p^*}{\partial y^*}. \tag{A-5b}$$

Together with the continuity equation, which can be derived as

$$\frac{\partial u^*}{\partial x^*} + \frac{\partial v^*}{\partial y^*} = 0, \tag{A-6}$$

we have three partial differential equations governing u^*, v^*, and p^*. Please do not be intimidated by them. We will let Bob solve them. With proper boundary conditions, Bob should be able to solve them if the value of Re is given, since they constitute a pure math problem.

It is sufficient for heat-transfer engineers to know that u^*, v^*, and p^* are functions of x^*, y^*, and Re, nothing else (unless some additional parameters appear in the boundary condition). If we are interested in, for example,

$$\left(\frac{\partial u^*}{\partial y^*}\right),$$

then this derivative remains a function of x^*, y^*, and Re. But if the wall gradient, i.e.,

$$\left(\frac{\partial u^*}{\partial y^*}\right)_{y^*=0},$$

is sought, then it should be a function of x^* and Re only. Furthermore, if we are interested in the average value of this wall gradient over , this final value

$$\left(\frac{\partial u^*}{\partial y^*}\right)_{y^*=0} \quad \text{should become a function of Re only. It can be readily shown that}$$

$\overline{C_f} = \overline{\tau_w}/(0.5\rho u_\infty^2)$ should be a function of Re only as well, without even solving the partial differential equations. Ndm is powerful.

A-2 Governing Equation for T

Starting from the first law of Thermodynamics for open systems,

$$\dot{m}_{in}\tilde{h}_{in} - \dot{m}_{out}\tilde{h}_{out} + Q_{in} + \dot{W}_{shaft,in} = \left(\frac{dU}{dt}\right)_{cv}$$

we can also derive, for steady-state problems,

$$u\frac{\partial T}{\partial x} + v\frac{\partial T}{\partial y} = \alpha\left(\frac{\partial^2 T}{\partial x^2} + \frac{\partial^2 T}{\partial y^2}\right) \tag{A-7}$$

by replacing u with T, with , and deleting the pressure-gradient term in Eq. (A-4b).

Lesson 12

Forced-Convection External Flows (III)

In this lesson, we will study how both the temperature inside the thermal boundary layer and Nu correlations can be found. If heat convection is the most important mode among three heat transfer modes, then finding Nu correlations may be the most important subject in heat convection.

Nomenclature

$c1$ = a convenient dimensionless quantity, defined as $(\Delta y)^2 \, pe/(HCL^*L_x)$ in a Matlab code

f = Blasius similarity function, defined as dimensionless stream function

HCL = heat convection length, defined as $[1.026+(Pr/0.7298)^{0.6}]^*L_x$, a purely empirical quantity

Nu_x = local Nusselt number, defined as $\bar{h}x/k$

Nu_L = Nusselt number, defined as hL/k

$\overline{Nu_L}$ = average Nusselt number over the entire plate, defined as $\bar{h}L/k$

$\overline{Nu_x}$ = average Nusselt number over [0, x], defined as $\bar{h}L/k$, but rarely used

Pr = Prandtl number, ν/α

T_s = temperature of the surface, in C; in K if there is radiation

η = Blasius similarity variable, defined in Eq. (1a)

θ = a dimensionless temperature, defined as $(T - T_s)/(T_\infty - T_s)$

1. Preliminary

In the soup problem presented in Lesson 10, we really did not officially solve a daily-life heat-convection problem. What we did was only to make a very crude approximation and obtain a qualitative answer.

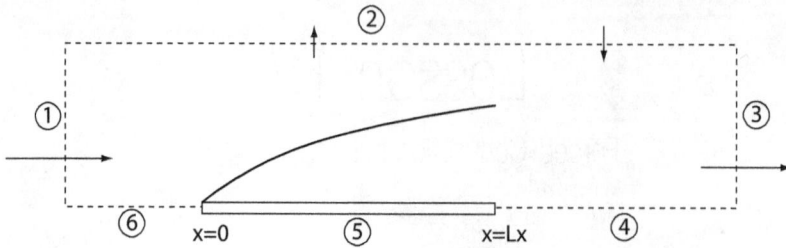

Fig. 12-1 Appropriate boundary conditions for the computational domain for boundary-layer flows

In a real heat-convection problem, if a certain degree of accuracy is strived for, we must:

(a) include in the energy-governing equation the y-direction convection term, $\rho c_p vT$,
(b) treat u as one of the field variables that are functions of x and y
(c) refine the numerical grid such that there are at least a few hundred grid nodes to capture the resolution of steep velocity and temperature gradients in the y direction.
(d) claim that we ignore the evaporation, radiation, and circumferential conduction loss.

A proper handling of this 2-D boundary-layer problem is believed to be like the one shown in Fig. 12-1. The computational domain is drawn in dashed lines.

On side 1: $u = u_\infty$, $T = T_\infty$, $v = 0$
On top face 2: $\partial u/\partial y = 0$, $\partial T/\partial y = 0$, $\partial v/\partial y = 0$. Note that in the wake region, most likely v_∞ is negative because of low pressures.
On side 3: $\partial u/\partial x = 0$, $\partial T/\partial x = 0$, $\partial v/\partial x = 0$. Such a specification of the outflow boundary condition is a subject of research in its own right.
On bottom face 4: $\partial u/\partial y = 0$, $\partial T/\partial y = 0$, $v = 0$, due to line of symmetry.
On the plate 5: $u = 0$, $v=0$, and $T = Ts$.
On the bottom face 6: same as on face 4.

Furthermore, $\partial p/\partial x$ is not necessarily small near the leading edge and the trailing edge. Assume that $nx \approx 100$ and $ny \approx 500$. At every node or computational cell, there are u, v, p, and T unknowns. Therefore, there are approximately 200,000 unknowns in total, and the same number of nonlinear algebraic equations that need to be solved.

Undoubtedly, if we must be involved with such computations, we will be easily overwhelmed by the great amount of computational work and may lose the opportunity of learning the physics of heat transfer.

Researchers in the past, including Blasius (1925), adopted the similarity formulation. Let us mention this procedure briefly, but tabulate the result in the appendix.

Blasius first introduced his similarity approximation containing the similarity variable defined as:

$$\eta = \frac{y}{x} Re_x^{1/2}, \text{ or } d\eta/dy = Re_x^{1/2}/x, \tag{1a}$$

and the dimensionless stream function defined as:

$$f = \psi / (v \, x \, u_\infty)^{1/2}. \tag{1b}$$

With Eqs.(1a, 1b), he was able to transform the original partial differential governing equations into a single 1-D nonlinear two-point boundary value problem, named Blasius equation:

$$f''' + 0.5\, f f'' = 0, \tag{2}$$

subject to: $f(0) = f'(0) = 0$, and $f'(8) = 1$

The prime indicates "differentiation with respect to η." A Matlab code solving Eq. (2) is given in the appendix.

A comparison of cubic-polynomial velocity profile and Blasius velocity profile is shown in Fig. 12-2. The red curve is the similarity solution.

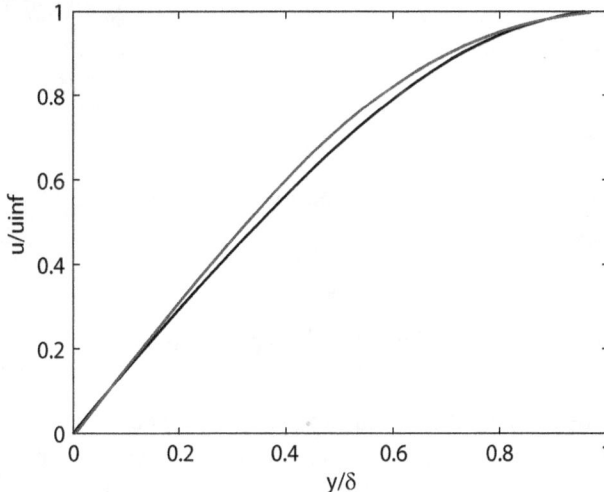

Fig 12-2 **Comparison of the cubic-polynomial profile and the similarity solution.**

At least two questions naturally arise: (i) How were two partial differential equations, governing u and v, transformed into a single ordinary differential equation (called Blasius equation), and (ii) how was Blasius equation solved?

The answer to (i) is in the 1980s, we would have gone over the related math in detail, because we had very few other options. In the 2010s, however, we have the access to powerful computers and software packages, and there are easier ways to solve the boundary-layer problem. Hence, we will skip the mathematical details. It is sufficient to state that the transformation is somewhat similar to the formulation of body-fitted coordinates, which is commonly used to treat irregular geometries. Advanced readers can refer to [1] for math details.

To answer question (ii), we can consult the literature, and will realize that Blasius adopted a certain power series to solve Eq. (2). That analysis constituted a part of his Ph. D. dissertation. Today, due to the advent of computers and numerical methods, the Blasius equation can be solved as a standard two-point boundary-value problem.

After Blasius published his work, scholars later also adopted his similarity approach and extended to more complicated problems. A notable one was reported in 1938, regarding flows over a flat plate not aligned with the direction of the free stream. Look up Wikipedia with the key word "Blasius." Further later on, scholars derived the energy equation similarly to the Blasius equation:

$$\theta'' + 0.5 Pr f \theta' = 0, \qquad (3)$$

subject to $\theta(0) = 0$ and $\theta(\eta_\infty) = 1$, where

$$\theta = (T - T_s)/(T_\infty - T_s)$$

and the value of η_∞ is in the neighborhood of 8, depending on the value of Pr.

3. Steps to Find Heat Flux at the Wall (from the Similarity Solution)

Steps to compute heat fluxes at the wall can be summarized below.

(a) Solve for $f(\eta)$ first and then $\theta(\eta)$. Note that f is a function of η only, not of any other parameters. Therefore, $f''(0)$ is just a pure number, and was found to be 0.332057. See the appendix.

(b) Since Pr number is imbedded in the θ governing equation, θ is a function of η and Pr. Hence, unlike $f''(0)$, the gradient $\theta'(0)$ is a function of Pr. In fact, it was found [2] to be

$$\theta'(0) = 0.332057 Pr^{1/3}. \tag{4}$$

It should be noted that, while Eq. (3) is a general governing equation valid for all Pr values, Eq. (4) is a correlation valid only for $0.6 \leq Pr \leq 15$. From this result, we further deduce

$$\text{(c)} \quad q''_s = -k \left(\frac{\partial T}{\partial y}\right)_{y=0} = -k \left(T_\infty - T_s\right) \left(\frac{d\eta}{dy}\right) \left(\frac{d\theta}{d\eta}\right)_{\eta=0}. \tag{5}$$

With $h = qs''/(T_s - T_\infty)$, the term $\left(\frac{d\eta}{dy}\right)$ given in Eq. (1a), and the term $\theta'(0)$ given in Eq. (4), we are able to obtain the final correlation given in the next section. Taking a little pause also alerts us with a general functional relationship for h as

$$h_{local} = \text{function of k, x, Re, and Pr.}$$

4. The Nu Correlation and Some Discussions

By definition, Nu is simply a dimensionless h. It is often convenient for us to write correlations in terms of all dimensionless parameters. Thus, Eq. (5) leads to

$$Nu_x = \frac{hx}{k} = 0.332\, Re_x^{1/2}\, Pr^{1/3} \tag{6a}$$

or, after taking the average, we obtain

$$\overline{Nu_L} = 0.664\, Re_L^{1/2}\, Pr^{1/3}. \tag{6b}$$

The Nu number is like a necktie wrapped in an elegant gift box. What is useful to the gift receiver is the necktie. But when we give the gift to him, the necktie itself does not look as presentable as the elegant box does. The heat transfer coefficient, h, is the necktie, which will actually be used to find the wall heat flux.

It is important for us to be aware that both Eqs. (6a) and (6b) are valid only for the following conditions:

(a) The wall temperature must be uniform, such that $\theta = (T - T_s)/(T_\infty - T_s)$ looks similar, and is a function of y/δ_T only, where δ_T is the thermal boundary layer thickness.

(b) The solid object must be a flat plate.

(c) There cannot be any suction or blowing on the plate, i.e., v at y=0 must be zero.

(d) The flow is laminar, i.e., $Re_L \lesssim$ 5e+5. There are different correlations for turbulent flows.

(e) The range of Pr must be $0.6 \leq Pr \leq 15$. If the fluid is, for example, liquid metal or motor oil, the correlations are different.

(f) The region must not be near the leading edge or the trailing edge of the plate.

For comparisons, what is the correlation for the soup problem given in Lesson 10? This problem will be left as an exercise. See Problem 12-2.

Prandtl number is an important parameter in heat convection. It indicates the relative strength between heat transfer and the viscosity effect. Figure 12-3 qualitatively depicts this trend. Motor oil flows are represented by the left figure; whereas liquid metal flows are represented by the right figure. Clearly, when Pr = 1, the two boundary-layer edges should coincide.

See Problem 12-10.

Equations (6a) and (6b) can be viewed almost as the Carnot cycle in Thermodynamics. Although highly idealized, they are useful in serving as a fundamental guide for us to develop other more realistic correlations.

In our lives, whenever starting a task or a project, we often need to develop a simple and idealized model first. The development of this model greatly helps us to understand the essentials, thus enabling us to focus on the improvements. Without XT's appearing in early 1980s, how can we have Pentium PC's today? It is also fair to say that, without Newton, there could have been no Einstein.

Equation (6b) states that Nu is a function of Re and Pr. If we care about only the general functional relationship, we actually do not need to go through such detailed analyses. We can arrive at the same conclusion by using Ndm only.

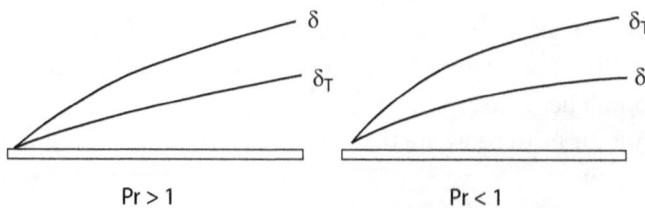

Fig. 12-3   Comparisons of thinkness of two boundary layers

In the boundary-layer momentum equation for u, there exist terms of momentum rate and shear stress,

$$\rho u^2 \propto \mu \frac{\Delta u}{\Delta y}, \text{ or } u_\infty^2 u^{*2} \propto \frac{\nu u_\infty}{L} \frac{\Delta u^*}{\Delta y^*} \text{ or } Re\, u^{*2} \propto \frac{\Delta u^*}{\Delta y^*}. \tag{7}$$

The symbol "$\propto$" indicates that only two representative terms are selected. Expression (7) is not an equation.

We now know with certainty that u* must be a function of x*, y*, and Re, without having to see the entire differential equation. Furthermore, looking at Fig. 12-4 showing grid node 2, we know that u_2* should be a function of x* and Re only, since y* has been fixed at j = 2. Next, we write

$$\tau_S = \mu \left(\frac{u_\infty}{L}\right) \frac{u_2^* - 0}{\Delta y^*} \quad \text{where "2" denotes "j = 2".}$$

Or,

$$C_f = \tau_S / (0.5\, \rho\, u_\infty^2) = \frac{2u_2^*}{Re\, \Delta y^*}.$$

Since u_2* is a function of only x* and Re, we conclude that C_f is also a function of only x* and Re, and that $\bar{C}_f$ is therefore a function of only Re (Δy^* is simply a grid interval chosen by us).

By similar logical arguments, we can also deduce that $\overline{Nu}_L$ should be a function of only Re and Pr. This deduction will be left as an exercise problem. See Problem 12-1.

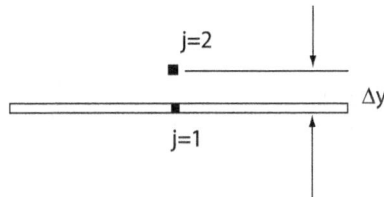

Fig. 12-4   A grid node very close to the wall

During the era of Informational Technology, it may be neither necessary nor practical for college engineering undergraduate and graduate students to learn the mathematical procedure required to obtain Eq. (6b). Instead, what they ought to know with confidence is

(a) In case of chickens and eggs, we may not know which one appeared first on Earth. In case of Nu correlations, for a given 2-D heat convection problem, we solve the governing equations, and therefore T(x,y) becomes available first. It then yields $(\partial T/\partial y)_{y=0}$, which subsequently yields the heat flux at the wall, which then yields h, which eventually yields Nu.
See Problem 12-7. See also the advantages of Ndm in Lesson 11.

(b) The logical path described in (a) must be kept in mind firmly. No, h is not originally derived from Eq. (6b).

(c) To learn a quick and approximate method to numerically obtain Eq. (6b) is very beneficial for us to understand the heat convection phenomenon.

6-1. A Quick and Approximate Method

Let us modify the governing equation for T_2 in the soup problem slightly, to account for a grid of multiple nodes, and an empirical dx, called Heat Convection Length (HCL) [3]. The modified equation takes the form of:

$$\Delta y\, u_\infty \left(\frac{-T_\infty + T_j}{HCL}\right) = \alpha \left(\frac{T_S - 2T_j + T_N}{\Delta y}\right), \text{ or}$$

$$c_1\left(-T_\infty + T_j\right) = T_S - 2T_j + T_N, \text{ where } c_1 = Pr\,Re\,\frac{(\Delta y)^2}{Lx\,HCL}.$$

In the Matlab code below, $T_S = 15C$ and $T_\infty = 45C$.

A recommended procedure to study this Matlab code can be

(a) cut and paste the code, run it, and see with our own eyes that the figure we obtain is the same as the one shown in Fig. 12-5.

(b) simplify the code to running only a single case: u = 10m/sec and Pr = 0.72978.

(c) change u = 10 m/sec (a fixed value) to a *for loop*, but keep Pr = 0.72978.

(d) change Pr = 0.72978 (a fixed value) to a *for loop*.

(e) For lower values or higher values of Pr, the agreement between the computed solution and the literature solution may not be good any more.

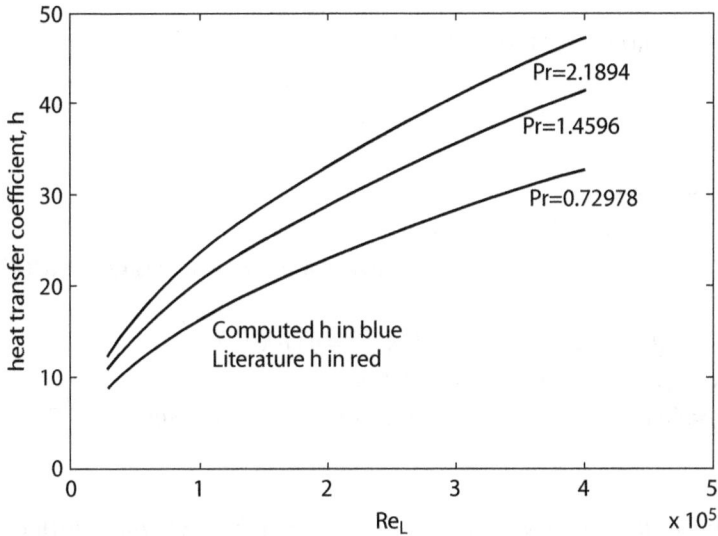

Fig. 12-5 Values of h as a function of Re_L parameterized in Pr

6-2. A Matlab code generating Nu correlation

```
clc; clear
% Most quantities here are dimensional.
Lx=.3;
ny=1000; Ly=Lx/5; dy=Ly/ny; dys=dy*dy; nyp=ny+1;
rhoa=1.18; cpa=1005; ka=.026; aLfa=ka/(rhoa*cpa); Tinf=45;
for iPr=1:3 % investigate cases for three Pr values
new=1.6e-5; Pr=iPr*new/aLfa;
HCL=(1.026+(Pr/0.7298)^0.6)*Lx;%The heat convection length was found after
many numerical
  % experiments, and is purely empirical.
  for iu=1:11 % Vary u_inf, thus vary Re_L.
u(iu)= 1.5 +(iu-1)*2;
Re(iu)=u(iu)*Lx/new; Pe=Pr*Re(iu);
c1=dys*Pe/(HCL*Lx); % a convenient term
for j=1:nyp; y(j)=(j-1)*dy; end
T(1)=15; T(nyp)=Tinf;
a=zeros(nyp, nyp);
% It is convenient to set all the elements in
% the coefficient matrix to zeros first.
%————————
a(1,1)=1; b(1)=T(1); a(nyp,nyp)=1; b(nyp)=T(nyp);
for j=2:ny
```

```
a(j,j-1)= -1; a(j,j)=(2+c1); a(j,j+1)= -1;
b(j)= c1*Tinf;
end
T=a\b';
%————————
Qs=ka*(1.5*T(1)-2*T(2)+.5*T(3))/dy;
% We can also use the simple 2-point approximation, which is Qs = ka*(T(1) – T(2))/
dy.
DT=T(1)-Tinf; % T(1)= Ts; Tinf = T(nyp) = T(1001)
h_cur(iu)=Qs/DT; % current computations
h_Lit(iu)=.664*Re(iu)^0.5*Pr^(1/3)*ka/Lx; % literature values
end
%
plot(Re,h_cur,Re,h_Lit,'r'); hold on % The two curves are almost indistinguishable.
xlabel('Re_L'); ylabel('heat transfer coefficient, h')
text(1.1e5, 15, 'Computed h in blue')
text(1.1e5, 13, 'Literature h in red')
text(3.47e5, h_Lit(9), ['Pr=',num2str(Pr)])
end
hold off
```

>>>>>>>>>>

We can run numerical experiments with the code to answer a few questions:

(a) As k increases, T tends to become more uniform, and $\left(\frac{\partial T}{\partial y}\right)_{y=0}$ tends to decrease.

So, will $q_s'' = -k\left(\frac{\partial T}{\partial y}\right)_{y=0}$ increase or decrease (which quantity, k or T gradient,

increases more rapidly)?
To give a semi-analytical answer, we realize: $q_s'' \propto k\,Pr^{1/3} \propto k\,k^{-1/3} \propto k^{2/3}$.
To give a numerical answer, we run the Matlab code.
(b) Is this T profile close to the cubic-polynomial approximation?
(c) We need to keep in mind that both the literature result and the current result
agree only in a certain parametric range.
(d) The wall heat flux is obtained by **heat conduction** at the wall. This value in turn
provides us with the h value. It is the author's observation that, during technical
discussions, some scholars would quickly sketch Fig. 12-6, stating, "We supply

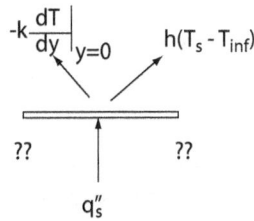

Fig 12-6   Conceptually incorrect energy balance over a plate

the energy to the plate, and the supplied energy is equal to the sum of heat conduction and heat convection...." This concept is thus fatally incorrect.

See Problem 12-6.

7. An Example Regarding Convection and Radiation Combined

Let us now study a real-life example below.

Example 12-1

Given: an air flow over a flat plate of Lx = 0.1m, T_∞ = 25C, T_s =500C, ε = 0.9, u_∞ = 8m/sec, Z = 1 m deep. Other values of properties are given in the Matlab code. The plate is facing the outer space (namely, no radiation arrives on the plate) and resting on the earth ground.

Find: how much energy (kW) is needed to supply to the plate in order to maintain Ts at 500C steadily.

```
clc; clear
Tinf=25; e=0.9; Ts= 500; uinf=8; Lx=0.1; Z=1;
new=1.6e-5; k=0.026; rho=1.1614; cp=1005; sig=5.67e-8;
Re = uinf*Lx/new % = 62500;
aLf = k/(rho*cp); Pr = new/aLf % = 0.7183
Nu = 0.664*Re^0.5*Pr^(1/3)% = 132.969
h = Nu*k/Lx % = 34.57 W/m^2-K
qconv = h*(Ts-Tinf)*Z*Lx % = 1.6422 kW
qrad = e*sig*(Ts+273)^4*Z*Lx % = 1.822 kW
q_supply = qconv + qrad % = 3.4642 kW
```

At Ts = 466.5C and u_∞ = 8 m/s, the radiative rate is approximately equal to the convective rate (1.526 kW)

See Problems 12-3, 12-4, and 12-5.

First, it is intuitive to us that the heat transfer effect between the freestream fluid and the wall decreases as x increases. Indeed, h decreases as x increases, confirming our intuition. The parameter, Nu_x, however, increases as x increases.

Second, when looking up a table of Pr values for various materials, we will find:

Pr of liquid metals ~ O (0.001), Pr of air ~ O (1).

Pr of water ~ O (1 to 10), Pr of motor oils ~ O (1000),

where O( ) denotes "order of magnitude." Intuitively, don't we expect the heat transfer effect by liquid metal flows to be better than motor oil flows? Yet, why is Nu ~ $Pr^{1/3}$? See Problem 12-8.

9. Brief Examination of Two More External Flows

Other than flows over flat plates, there can be various different flow situations. Let us mention only two representative ones below.

(a) In the stagnation corner flow shown in Fig. 12-7, $\partial p/\partial x$ or $\partial p/\partial y$ cannot be ignored, unlike in the boundary-layer flow case.

(b) In the underground storage enclosure as shown, we are interested in knowing how much energy should be supplied to keep T_storage at 15C. But where is the leading edge where x=0? Also, if the air temperature inside the enclosure is 15C, and if $T_\infty = -5C$, what is the temperature of the enclosure's top floor?

See Problem 12-9.

One of the key issues for both problems is that we do not know the value of h. We definitely cannot treat Eq. (6b) as a cookbook recipe, take it, and use it directly to compute these two problems. At this moment, are we capable of establishing nodal governing equations for u, v, p, T, and ρ, solving them simultaneously, and obtaining wall heat fluxes, thus h?

If we are, it is wonderful. If we are not, let us proceed to learn internal flows described in the next three lessons where we will be given the opportunity to learn simple numerical methods to tackle u, v, and p. Hopefully, this knowledge can be applied to external flows.

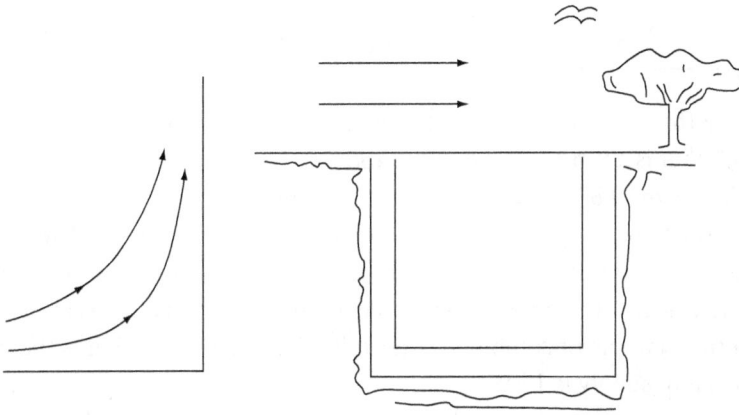

Fig 12-7 **Two representative cases that are different from idealized boundary-layer flows studied here**

10. Summary

This lesson, which may be the most import one in the entire book, can be summarized as follows:

(a) The boundary of the boundary-layer computational domain and the associated boundary condition are described.
(b) A classical approach of finding Nu is briefly described.
(c) An approximate and quick method that introduces the "heat convection length" is presented. The Matlab code can be used to perform numerical experiments.
(d) A combined convection and radiation problem is solved in an example.
(e) Two external flows that are not blown over flat plates are briefly examined.

11. References

1. Tien-Mo Shih, *Numerical Heat Transfer,* Springer-Verlag, 1984, ISBN 3-549-13051-9.
2. E. Pohlhausen, Der Warmeaustausch Zweschen Festen Korpen und Flussigkeit mit kleiner Reiburg und kleiner Zeit, *Z. Angew. Math. Mech.*, vol. 1, pp. 1–115, 1921.
3. Tien-Mo Shih, Chandrasekhar Thamire, and Yu-Guang Jiang, Heat convection length for boundary-layer flows, *International Comm. for Heat and Mass Transfer,* vol. 38, pp. 405–409, 2011.

12-1 Take the two representative terms in the energy equation. Proceed to prove $\overline{Nu_L}$ that is a function of only Re and Pr.

12-2 What is the $\overline{Nu_L}$ correlation for the soup problem?

12-3 Revisit the example regarding combined convection and radiation shown in Fig. 12-8, and plot q_conv versus Ts and q_rad versus Ts in the same figure, where 30C < Ts < 530C. Do you see these two curves intersect?

12-4 In the example of combined convection and radiation, what is the value of Ts in C if q_supply is 1 kW?

12-5 Let us modify problem #4 slightly. Instead of T_inf = 25C, now there is a hot airflow at T_inf = 200C blowing over the plate. The plate is exactly at the thermal equilibrium by itself without receiving any energy supply. Find Ts in C.

12-6 Find q"_conduction from the plate to the airflow in problem #5. Think carefully.

12-7 Consider an airflow over a flat plate. The following data are relevant:
u_{∞} = 10 m/s, Lx = 0.5m, Δy = 5e-5m, v = 5.2097e-4 m^2/s, k = 0.0364 W/m-K, Ts = 27C, and T_{∞} = 300C. Also, $\theta = (T - T_s)/(T_{\infty} - Ts)$. The applied mathematician, Bob, has given us his computational output, a small portion of which is shown below:

x*	0.4	0.5	0.6	(x* = x/Lx)
θ	0.00454	0.00405	0.00370	(at y* = y/L = 0.0001)

Find Nu_x and compare with Nu_x obtained from Eq. 12-6a.

12-8 When looking up a table of Pr values for various materials, we will find:
Pr of liquid metals ~ O (0.001) and Pr of motor oils ~ O (1000).
Also, from Eq. (12-6b), we note that Nu ~ $Pr^{1/3}$, provided that everything else remains unchanged. These two results seem to lead to the conclusion that Nu values of motor oils are much higher than those of liquid metals. Since Nu is an

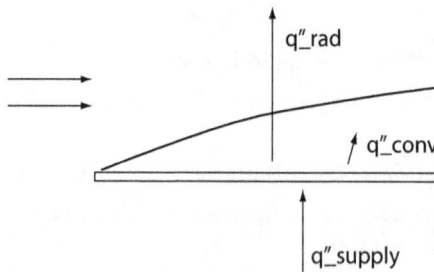

Fig. 12-8 **Combined convection and radiation over a flat plate**

indicator of heat convection effects, we may be tempted to think that the heat convection effect by motor oils is higher than that by liquid metals.

Intuitively, don't we expect the heat transfer effect to be the opposite? So, what is happening?

12-9 Revisit the soup problem in Lesson 10. Data required for the system schematic, shown in Fig. 12-9, are listed as follows:

k_soup = 0.61; k_air = 0.026; pe = 2; ε = .3; sig = 5.67e-8; dy = 0.01m; T(1) = 80C; T(5) = 20C;

The soup will radiate, but it receives no external radiation from the dining room (a bad assumption, but the problem is simplified). Also, now T(3) is not known, and the soup is so viscous that it will not circulate, allowing the assumption that 1-D steady state heat conduction takes place in the soup.

      (a) Find T(2), T(3), and T(4).

      (b) Find h.

Incidentally, this problem is known to be a conjugate problem, which involves both a convection domain and a conduction domain via the interface (node 3).

12-10 In liquid metal boundary-layer flows over a flat plate, which of the following statements makes the best sense?

(a) u is almost equal to u_∞ everywhere inside the thermal boundary layer.

(b) the conduction in y direction is very small, in comparison with the convection also in y direction.

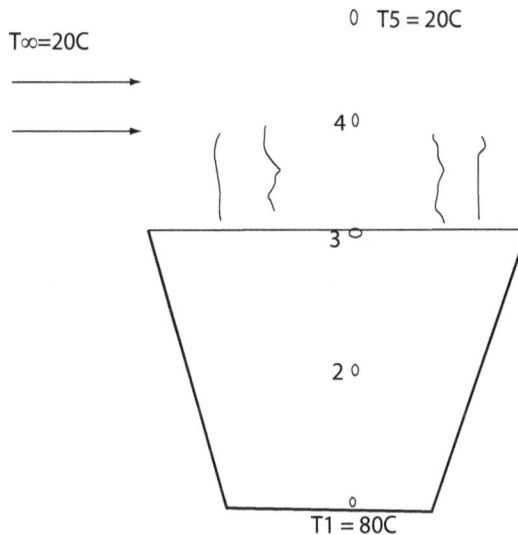

Fig. 12-9   The soup problem revisited. But now T(3) is not known. Hence the phenomenon constitutes a consjugate problem.

(c) the conduction in the streamwise flow direction is much greater than that in the transverse direction.

(d) none of the above.

---

13. Appendix [to find $f'(\eta)$ and the value of $f''(0)$]

Note that g is introduced, where $g = f'$, so that g'' can replace f''', and the Blasius equation can be reduced to a second-order nonlinear differential equation. Once g is computed, f can be found from the trapezoidal rule. In the computation, g and f must be solved simultaneously and iteratively. Fortunately, the solution has converged.

A-1. Matlab Code

```
clc; clear
% Choose Le=4.6407 for the eta domain for comparison of u profiles
% Choose Le > 8 to obtain f"(0) = 0.332.
Le = 10;
ne=2000; de=Le/ne; nep=ne+1;
for i=1:nep
 eta(i)=(i-1)*de; g(i)=0; f(i)=0; b(i)=0;
end
g(1)=0; g(nep)=1; % introduce g to be f'
for iter=1:12 % need to iterate because there is a nonlinear term
 a(1,1)=1; b(1)=0;
 for i=2:ne
 c1=.25*de*f(i);
 a(i,i-1)=1-c1; a(i,i)=-2; a(i,i+1)=1+c1;
 end
 a(nep,nep)=1; b(nep)=1;
 g=a\b';
 for i=2:ne
 f(i)=f(i-1)+.5*de*(g(i-1)+g(i)); % use trapezoidal rule
 end
 fppz=(-1.5*g(1)+2*g(2)-.5*g(3))/de; % 3-pt approximation
 %fprintf('%5.0f %9.4f \n', iter, fppz)% = 0.3398
end
for i=1:nep
 ksi(i)=eta(i)/Le;
```

```
u_cubic(i)=1.5*ksi(i)-0.5*(ksi(i))^3;
if(i==1); fpp(i)=fppz;
elseif(i==nep); fpp(i)=0;
else fpp(i)=(g(i+1)-g(i-1))/(2*de); end
end
% % plot(ksi, g, 'r', ksi, u_cubic); xlabel('y/\delta'); ylabel('u/uinf')
% % hold off
for i=1:nep
 fprintf('%9.3f %9.4f %9.4f %9.6f \n', eta(i), f(i), g(i), fpp(i))
end
```

A-2. Table of η, f, f', f'' Distributions

0.000	0.0000	0.0000	0.332057	0.135	0.0030	0.0448	0.332034
0.005	0.0000	0.0017	0.332057	0.140	0.0033	0.0465	0.332032
0.010	0.0000	0.0033	0.332057	0.145	0.0035	0.0481	0.332029
0.015	0.0000	0.0050	0.332057	0.150	0.0037	0.0498	0.332026
0.020	0.0001	0.0066	0.332057	0.155	0.0040	0.0515	0.332023
0.025	0.0001	0.0083	0.332057	0.160	0.0043	0.0531	0.332019
0.030	0.0001	0.0100	0.332057	0.165	0.0045	0.0548	0.332015
0.035	0.0002	0.0116	0.332056	0.170	0.0048	0.0564	0.332012
0.040	0.0003	0.0133	0.332056	0.175	0.0051	0.0581	0.332008
0.045	0.0003	0.0149	0.332056	0.180	0.0054	0.0598	0.332003
0.050	0.0004	0.0166	0.332056	0.185	0.0057	0.0614	0.331999
0.055	0.0005	0.0183	0.332055	0.190	0.0060	0.0631	0.331994
0.060	0.0006	0.0199	0.332055	0.195	0.0063	0.0647	0.331989
0.065	0.0007	0.0216	0.332054	0.200	0.0066	0.0664	0.331983
0.070	0.0008	0.0232	0.332054				
0.075	0.0009	0.0249	0.332053	0.250	0.0104	0.0830	0.331913
0.080	0.0011	0.0266	0.332052	0.300	0.0149	0.0996	0.331809
0.085	0.0012	0.0282	0.332051	0.350	0.0203	0.1162	0.331663
0.090	0.0013	0.0299	0.332050	0.400	0.0266	0.1328	0.331469
0.095	0.0015	0.0315	0.332049	0.450	0.0336	0.1493	0.331221
0.100	0.0017	0.0332	0.332048	0.500	0.0415	0.1659	0.330910
0.105	0.0018	0.0349	0.332046	0.550	0.0502	0.1824	0.330532
0.110	0.0020	0.0365	0.332045				
0.115	0.0022	0.0382	0.332043	0.600	0.0597	0.1989	0.330079
0.120	0.0024	0.0398	0.332041	0.650	0.0701	0.2154	0.329544
0.125	0.0026	0.0415	0.332039	0.700	0.0813	0.2319	0.328921
0.130	0.0028	0.0432	0.332037	0.750	0.0933	0.2483	0.328205

0.800	0.1061	0.2647	0.327389	3.200	1.5691	0.8761	0.139129
0.850	0.1198	0.2811	0.326466	3.400	1.7469	0.9018	0.117877
0.900	0.1342	0.2974	0.325432	3.600	1.9295	0.9233	0.098087
1.000	0.1656	0.3298	0.323007	3.800	2.1160	0.9411	0.080126
1.100	0.2002	0.3619	0.320071	4.000	2.3057	0.9555	0.064235
1.200	0.2379	0.3938	0.316589	4.200	2.4980	0.9670	0.050520
1.300	0.2789	0.4252	0.312528	4.400	2.6924	0.9759	0.038973
1.400	0.3230	0.4563	0.307865	4.600	2.8882	0.9827	0.029484
1.500	0.3701	0.4868	0.302580	4.800	3.0853	0.9878	0.021871
1.600	0.4203	0.5168	0.296663	5.000	3.2833	0.9915	0.015907
1.700	0.4735	0.5461	0.290111	5.500	3.7806	0.9969	0.006579
1.800	0.5295	0.5748	0.282931	6.000	4.2796	0.9990	0.002402
1.900	0.5884	0.6027	0.275136	6.500	4.7793	0.9997	0.000774
2.000	0.6500	0.6298	0.266751	7.000	5.2792	0.9999	0.000220
2.100	0.7143	0.6560	0.257809	7.500	5.7792	1.0000	0.000055
2.200	0.7812	0.6813	0.248351	8.000	6.2792	1.0000	0.000012
2.300	0.8505	0.7057	0.238426	8.500	6.7792	1.0000	0.000002
2.400	0.9223	0.7290	0.228092	9.000	7.2792	1.0000	0.000000
2.500	0.9963	0.7513	0.217412	9.500	7.7792	1.0000	0.000000
2.600	1.0725	0.7725	0.206455	10.000	8.2792	1.0000	0.000000
2.700	1.1508	0.7925	0.195294				
2.800	1.2310	0.8115	0.184007				
2.900	1.3130	0.8293	0.172670				
3.000	1.3968	0.8460	0.161361				

A-3 Brief Explanations of the Table

(a) The value of $f''(0) = 0.332057$ is a very valuable piece of information. It represents the dimensionless shear stress at the wall for a boundary-layer flow over a flat plate. Without exaggeration, it can be stated that the queen bee is the most essential individual in the entire hive. Similarly, $f''(0)$ is the most important result in the entire Blasius analysis, or entire boundary-layer flow analyses.

(b) From at the wall to $\eta \approx 0.2$, values of $f''(\eta)$ remain fairly constant, suggesting that, indeed, the assumption that u is linear in y near the wall (needed for obtaining the cubic-polynomial velocity profile) is valid.

(c) When η reaches 6, the value of u has already reached its asymptotic value, u_∞. Strictly speaking, for the engineering sense, computations beyond $\eta = 7$ are not necessary.

Lessons 13

Internal Flows (I)—Hydrodynamic Aspect

External flows have been studied in Lessons 10–12. The next topic for us to learn will logically be internal flows in channels, pipes, or tubes. Since parallel-plate channel flows in Cartesian coordinates are not involved with factors of r or 1/r, saving us much time from algebraic manipulations, we will focus on them. After learning the fundamental concepts, we can readily extend them to circular cross-sectional channels.

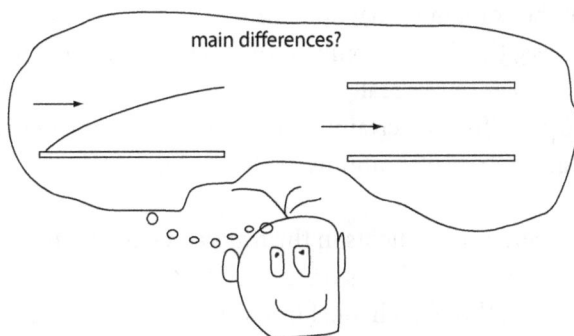

Nomenclature

b = half height of the channel, m

B = a positive dimensionless parameter defined in Eq. (6)

C_f = a dimensionless shear stress at the wall, $\tau_s/(0.5\rho u_m^2)$

D_h = hydraulic diameter defined in Eq. (1), m

Ndm = Nondimensionalization

Δp = pressure drop between the entrance and the exit of a channel

Re = Reynolds number, um*b/ν

um = mean streamwise velocity, m/s

Z = the depth of the channel, m

ξ = y/b, dimensionless y

A few characteristics can be identified regarding main differences between external flows and internal flows:

(a) pressure gradients in the streamwise direction

Inside channels, the flows are retarded by considerable frictional forces due to the presence of the confining walls. Eventually, blowers, fans, or pumps are needed to be installed to overcome these frictional forces, otherwise flows will simply stop. Even in the developing-flow region, dp/dx cannot be neglected, because the "free" stream is, strictly speaking, not free any more. The flow does "feel" the presence of the opposite plate, partly because the "free" stream velocity increases as the flow further moves into the channel.

See Problem 13-1.

(b) specification of the computational domain

In most external flows, the fluid is confined only by the plate on one side, with the other side open to the ambience. We really do not have any idea how far the influence of the plate can reach a priori. This uncertainty often makes the problems more challenging than cases where geometries of confining walls are clearly defined.

(c) velocity and temperature gradients in the transverse direction

In general, the wall velocity gradients or temperature gradients in boundary-layer flows are much higher than those in channel flows. The reasons are at least twofold: (i) the velocity is faster in the former, and (ii) the boundary-layer thickness is thinner than the height of the channel flow. Consequently, the convective terms in the governing equations are larger in magnitude. The nonlinearity is more severe in the former. See Problem 13-2.

2. Two Regimes (or Regions)

Two regimes of distinctive characteristics can be identified. Near the entrance of a channel, the flow gradually develops from a uniform velocity profile of an incoming flow to a parabolic velocity profile. See Fig. 13-1. Analyses of hydrodynamic and thermal aspects of this regime is also known as Graetz problems [1]. After this developing regime, the velocity profile of the flow remains unchanged along x. The regime is called hydrodynamically fully developed flow.

In the literature and in the industries, the hydraulic diameter, D_h, is extensively used. It is defined as

$$D_h = 4A_c/p. \tag{1}$$

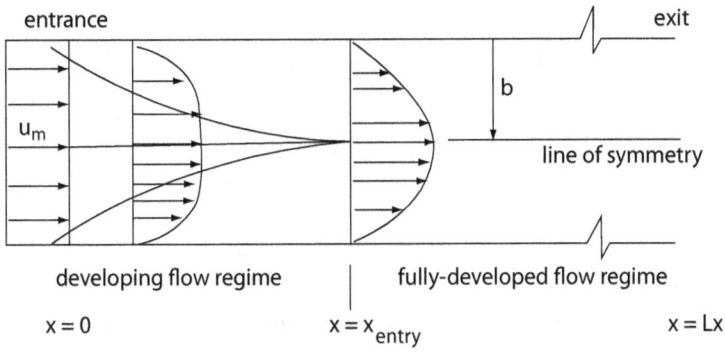

Fig. 13-1   Channel-flow schematic

It is so defined such that, for circular tubes,

$$D_h = 4(0.25\pi D^2)/(\pi D) = D,$$

the hydraulic diameter is exactly equal to the diameter of a tube. For two-parallel-plate channels shown in Fig. 13-2, using the definition in Eq. (1), we can find D_h to be

$$D_h = 4 (Z * 2b)/(4b + 2Z) \approx 4b, \text{ if } Z \gg b.$$

According to this definition, it was reported that the distance between the entrance and the position where two boundary-layer edges meet is approximately

$$x_{entry} \approx 0.05 \, D_h \, Re_D = 0.8 \, bRe. \tag{2}$$

See also Problem 13-3.

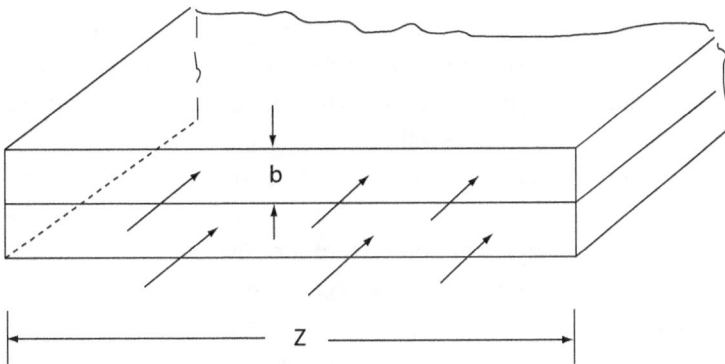

Fig 13-2   Sideview of the entrance of channel flows

In internal channel flows, there have been various definitions of Reynolds numbers, such as

$$u_m D/v,\ u_m R/v,\ u_c D/v,\ and\ u_c R/v,$$

where subscripts "c" and "m" stand for "centerline" and "mean", respectively. In this textbook, the definition, $Re = u_m b/v$, will be used, unless otherwise specified.

Also, only laminar flows will be considered in this lesson. Advanced readers who are interested in turbulent flows may refer to Lesson 17 and other heat transfer and fluid mechanics resources.

In general, for $Re_D \leq 2300$, the flows remain laminar.

The developing regime is somewhat similar to boundary-layer flows. However, now the flow domain is confined on both sides, and is thus better defined. Perhaps we can attempt to use a crude model to find u, v, and p qualitatively.

3. A Coarse Grid to Find u, v, and p in the Developing Regime

As in the soup problem in Lesson 10, a coarse grid is given in Fig. 13-3 for us to taste a little flavor of developing flows by finding u, v, and p without having to get too immersed in the numerical and programming complexities. We still can see qualitative trends and avoid complicated numerical methods to find hydrodynamic behaviors. When time permits in the future, we can then readily employ the same concept and increase the resolution of the grid.

The positions of u, v, and p are staggered so that the resulting coefficient matrix of unknowns represents a linearly independent system. (If u, v, and p are all located at the same grid node, there will be too many zero eigenvalues associated with the coefficient matrix. Readers are advised to look up journal papers regarding this issue, which will be omitted here).

There are only seven computational cells. The rightmost cell lies inside the fully developed region; therefore the velocity profile remains the same. Only half of the computational domain above the centerline is taken. The centerline constitutes the line of symmetry, on which v is zero everywhere.

To further simplify the problem, we will only examine low-Re flows, so that the magnitude of convection terms is small, and therefore these terms can be moved to the right-hand side of the equation and be temporarily treated as known values during iterations. For high-Re flows, these convection terms need to be linearized using Taylor's Series Expansion, and the algebraic manipulations become quite complicated.

In reference to Fig. 13-3, mass conservation over the computational cell 1, for example, yields

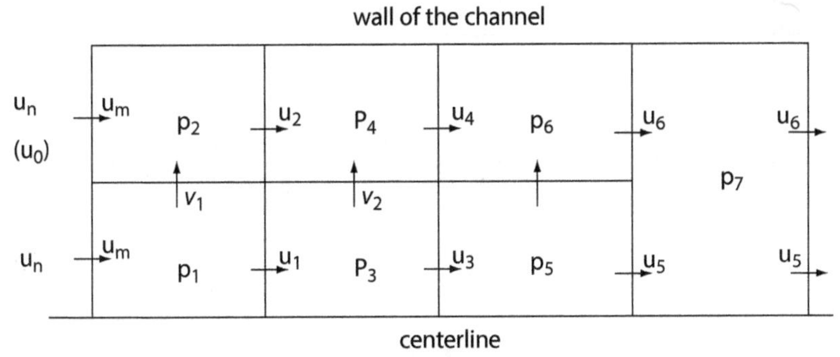

wall of the channel

centerline

Fig. 13-3   Grid system of seven computational cells

$(um - u2)^* \, \Delta y + v1^* \, \Delta x = 0.$

In this analysis, we need to carefully include grid intervals, Δx and Δy, in the equations, as they are not equal.

Momentum balance over the control volume surrounding u1 shown in Fig.13-4, for example, yields

$$(p1 - p3)^* \Delta y + \mu \left(\frac{u2 - u1}{\Delta y} \right) \Delta x = \rho u_1 \, (u_1 - u_0) \, \Delta y \,,$$

which can be rearranged into

$$\beta u1 - \beta u2 - p1 + p3 = \rho u_1 \, (u_0 - u_1),$$

where $\beta = \mu \Delta x / (\Delta y)^2.$ (3)

In considering the convection terms, we have neglected $\rho v \, \Delta x \, \partial u / \partial y$ for the purpose of simplicity. The derivation of other governing equations can be performed similarly. The Matlab code is given in the appendix. Let us present the converged numerical result graphically shown in Fig. 13-5 below:

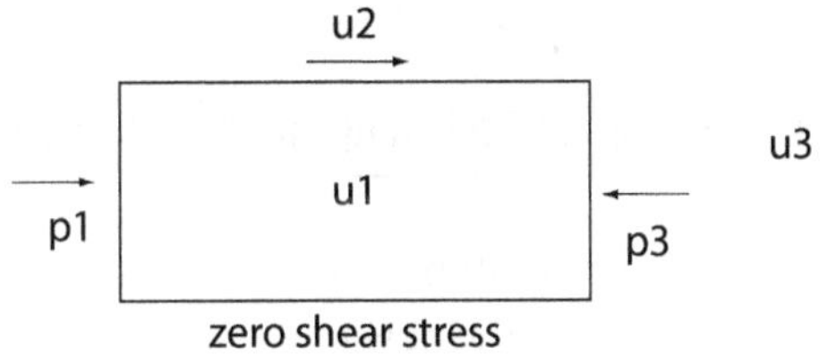

Fig. 13-4   Froce balance over the cell of u1

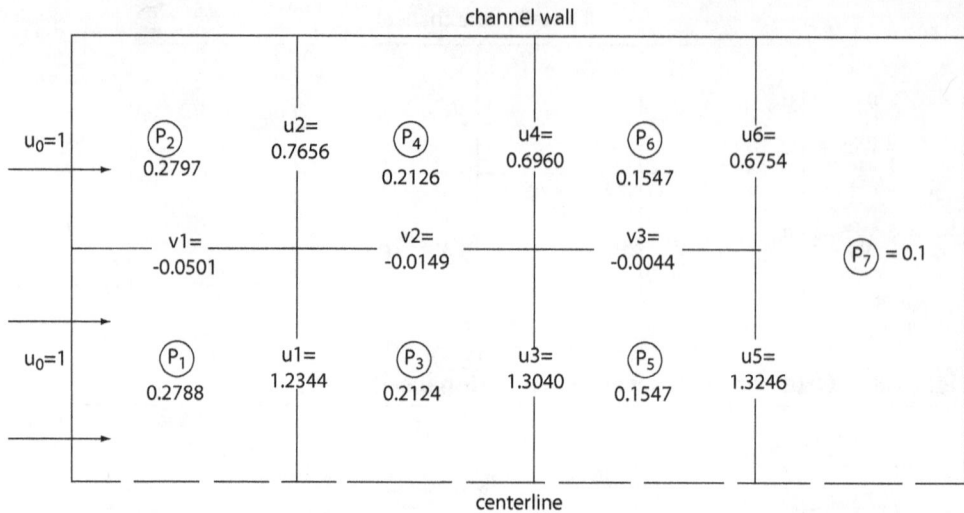

Fig 13-5 Solution of nodal values of u, v, and p in the developing regime. Note that those of u and v shown here are dimensionless quantities. True values should be divided by 100.

From the result, we can see that the trend looks qualitatively correct. For example,

(a) u1 + u2 = u3 + u4 = u5 + u6 = 2, suggesting that mass is conserved.
(b) v1 is negative, suggesting that u2 < um (or u_0) and the flow is retarded by the plate.
(c) v1 is negative, leading to that u1 > um, and the mass is conserved:
$$1*\Delta y + 0.0501 * \Delta x = 1.2344*\Delta y$$
(d) p2 is slightly greater than p1, pushing the flow toward the centerline.
(e) p1 is conspicuously greater than p3, pushing the flow into the channel.
(f) u values at y = 0.5Δy increase from 1 to 1.2344, to 1.3040, and to 1.3246, as they should.
(g) If we desire to include the energy equation, we can choose to locate nodal T's at u's.

See Problem 13-4.

4. An Analytical Procedure of Finding u(y) in the Fully Developed Regime

In the fully developed regime, momentum rates remain the same along x. Hence, the problem becomes much simpler than the developing-flow counterpart. It turns out that the problem can be even solved by using analytical methods.

There are at least two conditions under which the analytical method may be preferred to the numerical method:

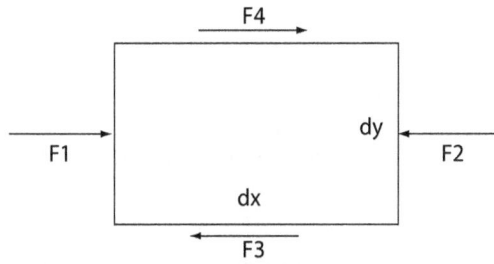

a differential control volume dx*dy*Z

Fig. 13-6   Force balance over dx*dy

(a) The analytical solution exists.

(b) The analytical procedure is not too algebraically complicated.

In the case of fully developed flows, these two conditions are actually met. Therefore, let us present the analytical procedure here, even though the numerical method is very simple, too.

The analytical procedure is available partly because we assume that

$$\partial u/\partial x = 0 \text{ and } v = 0$$

in the fully developed regime. As a result, there are no changes in the momentum rates, suggesting that all forces acting on the control volume are exactly balanced, as shown in Fig. 13-6.

In addition, the normal stress term vanishes as well, leaving us only pressure terms and shear stress terms, namely,

F1 = p dy,

F2 = (p dy) + $(\partial p/\partial x)dxdy$ by Taylor's Series Expansion,

F3 = $\mu (\partial u/\partial y)dx$,

F4 = $\mu (\partial u/\partial y)dx + \{\partial[\mu (\partial u/\partial y)]/\partial y\}dxdy$ also by Taylor's Series Expansion.

Since F1 − F2 − F3 + F4 = 0, and if μ = constant, we can readily obtain

$$\mu \frac{\partial^2 u}{\partial y^2} = \frac{\partial p}{\partial x} . \tag{4}$$

Before attempting to solve Eq. (4) for u(x,y), we can take a closer look to see if we can make further simplifications. Pleasantly, it has dawned upon us that

(a) $\frac{\partial p}{\partial x}$ cannot be a function of y, because v is equal to zero and hence $\partial p/\partial y = 0$.

(b) Neither can $\frac{\partial p}{\partial x}$ be a function of x, because, in terms of Eq. (1), we have learned

that u is not a function of x by the very definition of fully developed flows.

(c) This pressure gradient cannot be equal to zero, because otherwise there will be no force to drive the flow through the channel.

Based on (a) – (c), we conclude that $\frac{\partial p}{\partial x}$ must be a non-zero constant, implying we can proceed to integrate u with respect to y once and twice with ease.

Incidentally, we can also argue that the shear stress term cannot be a function of x, and the pressure gradient term cannot be a function of y. If so, then Y(y) = X(x) must be equal to a constant with no other choices.

See Problem 13-6.

With even more ease, let us first Ndm Eq. (4) into

$$\left(\frac{d^2 u^*}{d\xi^2}\right) = -B, \tag{5}$$

where $B = -\frac{b^2}{\mu u_m}\frac{\partial p}{\partial x}$, $u^* = u/u_m$, and $\xi = y/b$. $\qquad$ (6)

This parameter, B, is dimensionless, and its value is positive (because as x increases, p should decrease).

Integrating Eq. (5) once and twice, knowing that,
at $\xi=0$, $du^*/d\xi = 0$ and at $\xi=1$, $u^*=0$,
we obtain

$$u^* = 0.5\, B(1 - \xi^2). \tag{7}$$

At this moment, let us take a small pause and ask ourselves a question. Does Eq. (7) imply that u^* is a function of B and ξ? Or perhaps we can rephrase our question to be "Can B be assigned with any value, such as 1, 10, or 100 arbitrarily?"

The answer is imbedded in the definition of u^* (u/u_m), which dictates that the u^* value must vary between 0 and uc/um. Had B value been, say, 100, it is obvious that, at $\xi=0$, u^* is equal to 50, which states uc=50*um. This situation is highly unlikely. So, we are provided the hint that there must be a constraint for the B value.

By definition,

$b\, u_m = \int_0^b u\, dy$, which can be quickly transformed into

$1 = \int_0^1 u^*\, d\xi.$

Substituting Eq. (7) into the equation above leads to B = 3. Hence, Eq. (7) becomes
$u^* = 1.5\,(1 - \xi^2).$ $\qquad$ (8a)

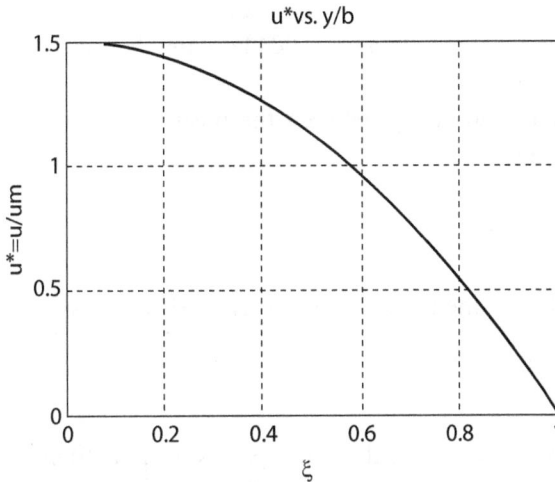

Fig. 13-7   Streamwise flow velocity versus y/b

Figure 13-7 plots Eq. (8a). It can be seen that the velocity located at $\xi \approx 0.58$ is equal to u_m.

See Problem 13-7.

Through our analysis, we have accomplished the following results:

(a) The velocity profile shown in Eq. (8a) is parabolic in the fully developed regime. It cannot be a cubic-polynomial profile because, if it were, dp/dx would have been a function of y. Neither can it be linear because, if it were linear, dp/dx would have been zero.

(b) From Eq. (8a), we have learned that

$u^* = u_c/u_m = 1.5.$ (2 for circular tubes).   (8b)

This information is quite useful. It can be used to calibrate experimental apparatus and instrumentation, and to validate computational fluid dynamics (CFD) computer codes.

There are not too many simple and well-defined benchmark problems in the area of fluid mechanics and heat transfer like this channel flow. In this sense, Eq. (8b) is very valuable for the academia and the industries. If someone claims that his muscles are very powerful, we can immediately challenge him to lift up a 200-pound weight for demonstration. By the same token, if someone claims that his CFD code is very powerful, we can immediately ask him to verify if it can produce the

computational result of u* = 1.4998 or 1.5004 far enough downstream in a channel for any Re < 2300/4.

(c) There exists a relationship between the mean velocity and the pressure drop. From Eq. (6), we obtain

$$u_m = -\frac{b^2}{3\mu}\frac{\partial p}{\partial x} = \frac{b^2}{3\mu}\frac{\Delta p}{L} \ . \tag{8c}$$

(d) Most importantly, the dimensionless surface shear stress can be found to be

$$C_f = \tau_S/0.5\rho u_m^2 = \frac{6}{Re} \ . \tag{8d}$$

Incidentally, if the flow is turbulent, Eq. (8d) is invalid. Then the Moody chart [2] can be consulted to obtain C_f as a function of Re and surface roughness.

5. Application of the Results

Some applications can be examined.

Example 13-1

Given: a water flow in a channel of two parallel plates, L = 0.2m, Δp = 3.75 Pa, ρ = 1000 kg/m^3, ν = 8.55 e-7m^2/sec, and b = 0.0025m

Find: u_m

Sol:

$$u_m = \frac{b^2}{3\mu}\frac{\Delta p}{L} = \frac{(0.0025)^2}{3*(1000*8.55e-7)}\frac{3.75}{0.2} = 0.0457.$$

We may be curious to know if, indeed, the pressure-drop force across the channel of length L is balanced by the surface shear-stress force, assuming the whole channel is filled with a fully developed flow.

First, from Eq. (8c), we obtain

$$\Delta p \,(b\,Z) = 3\,\mu\,L\,Z\,u_m/b. \tag{9a}$$

On the other hand,

$$\tau_s\,(L\,Z) = C_f(0.5\,\rho\,u_m^2)LZ = 3\,\mu\,L\,Z\,u_m/b. \tag{9b}$$

Therefore, $\Delta p \, b = \tau_s \, L$, as expected.

The pressure force difference given in Eq. (9a) is seen to be proportional to μ, L, and u_m. This relationship is somewhat expected. As b increases, the cross-sectional area increases, and we seem to need more pressure power to drive the flow. But Eq. (9a) also shows that the pressure force difference is inversely proportional to b. This trend is interesting, suggesting that, as b increases, the decrease in pressure power is primarily due to the decrease in shear stresses.

6. Which Value Should We Use?

In Lesson 10, we found the shear stress on the flat plate. In this lesson, we also found the shear stress at the wall in Eq. (8d) or (9b). Then, at the location where two boundary-layer edges meet (or at the end of the developing flow) or at the beginning of the fully developed flow, which value should we use for τ_s?

From the boundary-layer analysis and the cubic-polynomial velocity profile, we have

$$\tau_s = \mu \left(\frac{\partial u}{\partial y}\right)_{y=0} = \frac{\mu u_\infty}{\delta} \left(\frac{du^*}{d\xi}\right)_{\xi=0} = 1.5 \, \frac{\mu u_\infty}{\delta} = 1.5 \, \frac{\mu u_c}{b} \; . \tag{10a}$$

From Eq. (9b), we obtain

$$\tau_s = 3 \, \frac{\mu u_m}{b} = 2 \, \frac{\mu u_c}{b} \; . \tag{10b}$$

The results in Eqs. (10a) and (10b), unfortunately, do not agree well. So, where did we go wrong?

If we recall our boundary-layer analysis carefully, the free stream velocity is u_∞, which must be uniform. If u_∞ is not uniform, the x-direction momentum governing equation (which should include dp/dx term) and all the results should become different. In the channel flow case, indeed, u_∞ increases from u_m to u_c within the developing

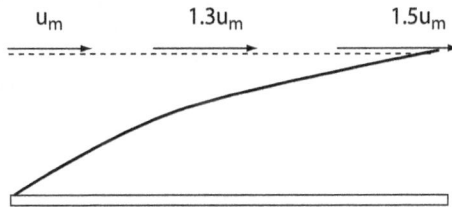

u_m $1.3u_m$ $1.5u_m$

Fig. 13-8   Acceleration of free stream velocities

regime, as shown in Fig. 13-8. So, u_∞ is definitely not uniform, and consequently Eq. (10a) should not be used.

See Problem 13-8.

7. Ndm and Parameter Dependence

Equation (8a) suggests that u* is a function of ξ only. Hence, $(du^*/d\xi)_{\xi=1}$ is just a pure number, depending on nothing. Subsequently,

$$\tau_S = \mu \left(\frac{\partial u}{\partial y}\right)_{y=b} = \frac{\mu u_m}{b} \left(\frac{du^*}{d\xi}\right)_{\xi=1}$$

depends on μ, u_m, and b. But

$$C_f = \tau_S/0.5\rho u_m^2 = (2/\text{Re}) \left(\frac{du^*}{d\xi}\right)_{\xi=1}$$

depends on only Re.

Somewhat surprisingly, we will find that Nu for thermally fully developed flows is a pure number, and depends on nothing, if the surface heat flux is kept constant. In Lessons 14 and 15, we will turn our attention to this topic.

8. Summary

This lesson can be summarized as follows:

(a) Both methods and solutions in the hydrodynamically developing and fully developed regimes have been presented. In the former, we used a coarse grid to find u, v, and p. In the latter, we found the u distribution analytically.

(b) Results have been applied to global momentum balance of entire channels.

(c) The difference between the developing regime and the boundary-layer flow regime has been briefly discussed.

9. References

1. Randall F. Barron, Xianming Wang, Timothy A. Ameel, and Robert O. Warrington, The Graetz problem extended to slip-flow, Int. J. Heat Mass Transfer, pp. 1817-1823, 1997.

2. Wikipedia.

13-1 In reference to Fig. 13-9, estimate and compare the pressure gradients, dp/dx, both inside the boundary layer flow and inside the developing channel flow.

13-2 Compare two typical values of velocity gradients both at y=b inside the fully developed channel airflow and at y=0 inside the boundary-layer airflow, to see if it is true that, indeed, the latter is greater.

13-3 Under the same values of b and u_m, estimate the ratio of x_{entry} for water flows to that for air flows inside the developing regime of channel flows.

13-4 Print out u1, u2, v2, u4, p2, and p4, and check if these values satisfy the continuity equation for the cell #4, and the x-direction momentum equation for the cell surrounding u2. Watch out the fact that the printed out u and v values are normalized on un for clarity.

13-5 It is important for us to understand the concept of nonlinearity, especially when we have entered the topic of heat convection, and are about to enter the topic of radiation. More specifically, what does it mean when we state, "we may be safe if the nonlinearity is small, otherwise the solution may diverge"? Try to answer this question by running the following problem:

$$3x + 2y = 5$$
$$x - y = c_1 (x^4 - xy)$$

c1=0.05;
xb=3; yb=4; % Use this initial guess.

(a) Iterate 10 times. The exact solution is x=y=1. Leave the nonlinear term on the right-hand side of the equation. Write the code yourself. Do not use the built-in functions in your magic calculators or Matlab.

If c_1 =0.05, the solution will converge. If c_1 =1, the solution will diverge. The nonlinear term, x^4, signifies a radiation term; xy signifies a convection term.

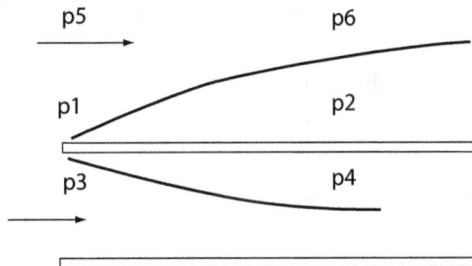

Fig. 13-9   Estimate dp/dx in two regimes

Is "love given by us to our beloved ones" linear with "love received by us from them?

(b) If $c_1 = 1$, how are we able to make the solution converge? Just comment on this.

13-6a If $X(x) = \sin(x)$ and $Y(y) = y^2$, and if $X = Y$, why must "$X = Y = $ constant" be true? For example, why can't we have:

$\sin(x) = y^2 = y - 0.334$, where $x = 1.2$ and $y = 0.9654$?

Comment on this.

13-6b In the hydrodynamically fully developed regime, the governing equation is

$$\mu \frac{\partial^2 u}{\partial y^2} = \frac{\partial p}{\partial x},$$

and we still need one more equation to determine values of p. There are several ways to cook an egg. Surely, this equation can be solved using other methods different from the analytical method, such as a numerical method. Outline this alternative step by step. Each step contains not more than two sentences. Take ny=40, b=0.01, and um=0.02 for a water flow.

13-7 Compare the shear stresses of the following four cases:
(a) pushing a brick resting on a wooden table at a constant slow speed
(b) a shear stress = hydrostatic pressure 1m beneath the water surface in a swimming pool
(c) water flowing in a channel
(d) air flowing over a flat plate

Relevant data are given below:

% for brick
cof=0.2; % coeff. of friction between a brick and a wood table surface (Ff = cof*W_brick)
mass_brick=1.5; g=9.8; A_brick=0.02;

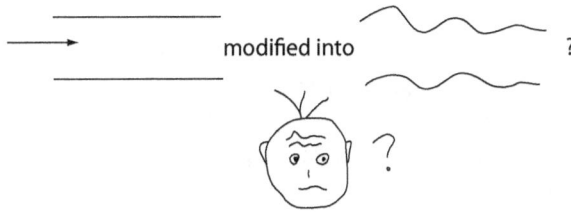

modified into ?

```
%
% for water in a channel
new=8.55e-7; rho=1000; mu=new*rho;
um=.04; b=0.01;
re=um*b/new % slightly below Re_critical.
%
% for air over a flat plate
new=1.6e-5; uinf=10; Lx=0.64; rho=1.1614;
re=uinf*Lx/new % slightly below Re_critical
```

13-8 How can we possibly modify a channel such that the pressure at the centerline in the developing regime remains approximately constant?

11. Appendix: Finding u, v and p in the Developing Regime

```
clc; clear % Plotting of u/um vs. y/b
%
Lk=1; nk=30; dk=Lk/nk; nkp=nk+1;
for i=1:nkp
 ksi(i)=(i-1)*dk;
 us(i)=1.5*(1-ksi(i)^2);
end
plot(ksi, us); xlabel('\xi'); ylabel('u*=u/um'); grid on
```

>>>>>>>>>>

```
clc; clear % Finding u, v, and p in the developing regime
%
b1=.001; % channel geometries (I), safer not to use the symbol "b" as the half height
nu=8.55e-7; rho=1000; mu=nu*rho; % fluid properties
un=.01; re=un*b1/nu; % flow conditions
pn= 0.1; % arbitrary reference pressure
Lx=0.8*b1*re; dx=Lx/4; dy=b1/2; dys=dy*dy; % channel geometries (II)
```

```
beta=mu*dx/dys;
%u1=q1; u2=q2; u3=q3; u4=q4; u5=q5; u6=q6;
%v2=q7; v2=q8; v3=q9;
%p1=q10; p2=q11; p3=q12; p4=q13; p5=q14; p6=q15;
a(15,15)=0; b(15)=0;
for i=1:6; u(i)=un; end;
for iter=1:20 % the solution has converged fortunately
a(1,1)= beta; a(1,2)=- beta; a(1,10)=-1; a(1,12)=1; % u1 eqn
a(2,1)=-beta; a(2,2)= 3*beta; a(2,11)=-1; a(2,13)=1; % u2 eqn
a(3,3)= beta; a(3,4)=- beta; a(3,12)=-1; a(3,14)=1; % u3 eqn
a(4,3)=-beta; a(4,4)= 3*beta; a(4,13)=-1; a(4,15)=1; % u4 eqn
a(5,5)= beta; a(5,6)=-beta; a(5,14)=-1; % u5 eqn
a(6,5)=-beta; a(6,6)= 3*beta; a(6,15)=-1; % u6 eqn
a(7,7)=mu/dy; a(7,10)=-1; a(7,11)=1; % v1 eqn
a(8,8)=mu/dy; a(8,12)=-1; a(8,13)=1; % v2 eqn
a(9,9)=mu/dy; a(9,14)=-1; a(9,15)=1; % v3 eqn
a(10,1)=-dy; a(10,7)=-dx; b(10)=-un*dy; % p1 eqn (mass conservation)
a(11,2)=-dy; a(11,7)= dx; b(11)=-un*dy; % p2 eqn (mass conservation)
a(12,1)=dy; a(12,3)=-dy; a(12,8)=-dx; % p3 eqn (mass conservation)
a(13,2)=dy; a(13,4)=-dy; a(13,8)=dx; % p4 eqn (mass conservation)
a(14,3)=dy; a(14,5)=-dy; a(14,9)=-dx; % p5 eqn (mass conservation)
a(15,4)=dy; a(15,6)=-dy; a(15,9)=dx; % p6 eqn (mass conservation)
b(1)=rho*u(1)*(un-u(1)); b(2)=rho*u(2)*(un-u(2));
b(3)=rho*u(3)*(u(1)-u(3)); b(4)=rho*u(4)*(u(2)-u(4));
b(5)=rho*u(5)*(u(3)-u(5))-pn; b(6)=rho*u(6)*(u(4)-u(6))-pn;

%
q = a\b'; u =q(1:6);
u_and_v = q(1:9)'/un % velocity solution normalized on un
end
p = q(10:15)'

% >>>>>>>>>>
u = 1.2344      0.7656      1.3040      0.6960      1.3246      0.6754

v = -0.0501     -0.0149     -0.0044

p = 0.2788      0.2797      0.2124      0.2126      0.1547      0.1547
```

Lessons 14
Internal Flows (II)—Thermal Aspect

In this relatively short lesson, let us turn our attention to the thermal aspect of channel flows. There is some groundwork that should be laid first.

Nomenclature

F1 = a fact in the first generation

H = a dimensionless parameter, defined in Problem 14-2, related to the case of constant Ts.

TFDF = an acronym standing for thermally fully developed flows

Tm = mean temperature, defined in Eq. (1), C or K

Ts = surface temperature, or temperature of the plate, C or K

α = thermal diffusivity, defined as $k/(\rho c_p)$, m^2/s,

θ = a dimensionless temperature, defined in Eq. (2)

ξ = dimensionless y, y/b

---

1. Definition of Tm

It is beneficial for us to pay special attention to the definition of Tm, which is different from how u_m was defined. Usually, for a given function $\varphi(y)$, we simply define its mean value to be

$$\varphi_m = \int_0^b \varphi \, dy/b \,.$$

For example, $u_m = \int_0^b u \, dy/b$.

For T_m, however, we establish its definition based on the enthalpy carried by the fluid. If there are three chunks of fluid, moving at u1, u2, and u3, each carrying $c_p T_1$, $c_p T_2$, and $c_p T_3$, respectively, then we define

Table 14-1	Calculations of u_m and T_m		
	fluid chunk 1	fluid chunk 2	mean
T	20C	40C	25C
u	3 m/s	1 m/s	2 m/s

$u_m T_m = (u_1 T_1 + u_2 T_2 + u_3 T_3)/3$.

Therefore, when applying this concept to the continuum sense, we have

$u_m T_m = \int_0^b u \, T \, dy/b$. (1)

Table 14-1 illustrates the calculation for T_m for two chunks of fluid. Note that the value of T_m is 25C, not 30C that is the simple arithmetic mean. It is important for us to have a clear recognition of this definition first in order to conduct further analyses correctly later.

2. Definition of Thermally Fully Developed Flows

Unlike the definition of hydrodynamically fully developed flows, thermally fully developed flows (TFDF) do not simply follow

$\partial T/\partial x = 0$.

For the assumption above to be true, most likely there is no energy interaction between the fluid and the channel walls, rendering the problem trivial and not too interesting. Instead, the condition of TFDF is [1]

$v = 0$ and $\partial \theta/\partial x = 0$, where

$\theta(\xi) = (T - T_s)/(T_m - T_s)$ (2)

is a function of ξ only. See Fig. 14-1 for sketched profiles of $T(x, y)$ and $\theta(\xi)$. Equation (2) intends to claim that, while T, T_s, and T_m all can be functions of x, the dimensionless form, θ, can be possibly independent of x. Hence, differentiating Eq. (2) with respect to x leads to

$\left(\dfrac{dT_m}{dx} - \dfrac{dT_s}{dx}\right)\theta = \left(\dfrac{dT}{dx} - \dfrac{dT_s}{dx}\right)$ (2a)

and with respect to y leads to

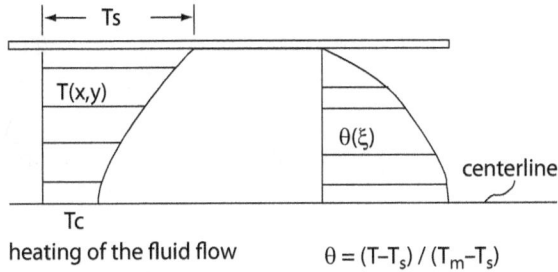

heating of the fluid flow $\theta = (T-T_s)/(T_m-T_s)$

Fig 14-1 **Typical profiles of dimensional temperature and dimensionless temperature**

$$(T_m-T_s)\frac{d\theta}{d\xi} = b\frac{\partial T}{\partial y} \tag{2b}$$

which can be used in the next section.

3. Justification of $\partial T/\partial x =$ constant

The rationale of assuming TFDF is that under this assumption, we may be able to analytically solve the energy equation

$$u\frac{\partial T}{\partial x} = \alpha\frac{\partial^2 T}{\partial y^2} \tag{3}$$

provided that a certain boundary condition on the wall is imposed. Let us examine the case of uniform q_s'' first, as it turns out to be easier than the case of uniform T_s.

Attempting to justify the constancy of $\partial T/\partial x$ can be a challenging logical process. Using Eq. (2b), we obtain

$$q_S'' = -k\left(\frac{\partial T}{\partial y}\right)_{y-b} = -\frac{k(T_m-T_s)}{b}\left(\frac{d\theta}{d\xi}\right)_{\xi=1} \tag{4}$$

In the equation above, q_s'', b, and k are constants, and

$$\left(\frac{d\theta}{d\xi}\right)_{\xi-1} \text{ is a pure number.} \tag{4a}$$

Therefore,

$(T_m - T_s)$ must be a constant, $\tag{4b}$

implying the left-hand side of Eq. (2a) must be equal to zero, further implying

$$\frac{dT}{dx} = \frac{dT_s}{dx} = \frac{dT_m}{dx} \tag{5}$$

At this moment, we have not yet succeeded in proving that dT/dx must be a constant. For example, $T = T_s = T_m = \sin(x)$ can also satisfy Eq. (5). Therefore, we are still in need of at least one more step.

In reference to Fig. 14-2, the first law of thermodynamics for open systems dictates

$$q_1 - q_2 - q_3 = 0 \tag{6}$$

where q1, q2, and q3 are three major convective energy components crossing the boundary of the control volume b Z Δx, and can be expressed by

$$q_1 = c_p\, b\, Z\, u_m\, T_m, \quad q_2 = q_1 + (dx) dq_1/dx, \text{ and}$$

$$q_3 = dx\, Z\, q_s'',$$

leading Eq. (6) to

$$\rho\, c_p\, b\, u_m \frac{dT_m}{dx} = -q_s'' \tag{7}$$

Note that heat conduction in the x direction has been neglected according to our assumptions that the axial heat conduction is much smaller than the transverse heat conduction, as shown in Eq. (3). Equation (7) offers us at least two important conclusions:

(a) If the flow is being heated up by the plate, q_s'' is negative, implying the dT_m/dx is positive. If the flow is being cooled down by the plate, q_s'' is positive, implying dT_m/dx is negative.
(b) If q_s'' is constant, then dT_m/dx must be a constant.

With the aid of conclusion (b), Eq. (5) now becomes

$$\frac{dT}{dx} = \frac{dT_S}{dx} = \frac{dT_m}{dx} = \text{constant.} \tag{8}$$

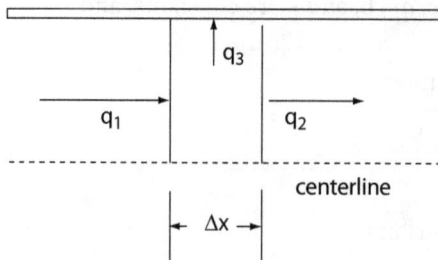

Fig 14-2   Apply the first law of thermodynamics to a finite control volume

The justification of Eq. (8), described in the previous section, illustrates a logical problem consisting of a series of steps that can be summarized in a form of genetic chart below. First, let us be entertained with a simple daily-life example:

Fact 1a = We study together in Maryland State. (a 1st-generation fact)
Fact 1b = Maryland State is situated in the U.S. (a 1st-generation fact)
Fact 1c = People who have studied together on the same campus are called alumni after graduation (a 1st-generation fact)

Hence,

F1a + F1b → F2a = We study together in the U.S. (a 2nd-generation fact can be produced)
F1a + F1c → F2b = We will be alumni after graduation and forever (another 2nd-generation fact)
Back to our channel-flow problem, we have:

F1a = Eq. (2), definition of θ
F1b = qs" is a constant
F1c = Eq. (7), energy balance over a segment of dx

These three facts are the original ones and can be considered as the 1st-generation ancestors. They will produce 2nd-generation facts:

F1a → F2a [Eq. (2a)]
F1a → F2b [Eq. (2b)]
F1a → F2c [Eq. (4a)]
F1b → F2d [Eq. (4)]

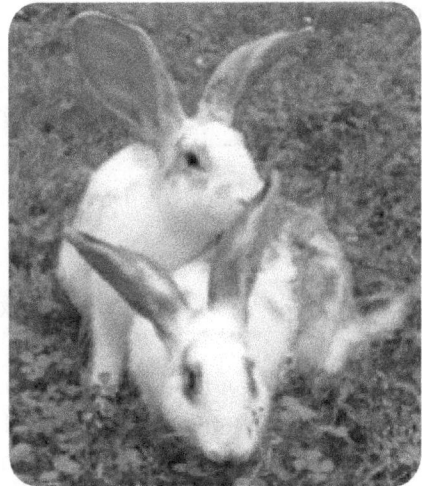

These four 2nd-generation facts then produce a 3rd-generation fact:

F2b + F2c + F2d → F3 [Eq. (4b)]

Subsequently, a 4th-generation fact is produced by:

F2a + F3 → F4 [Eq. (5)]

Finally, a 5th-generation fact is produced:

F1c + F4 → F5 [dT/dx is a constant]

This exercise allows us to gain heat-transfer knowledge as well as to cultivate our logical thinking ability. Needless to say, it is a very important and beneficial exercise. In order to fully understand this exercise, we are recommended to print out hard copies. Do not scroll our computer screens back and forth, trying to see what those equations are. Our heads will become dizzy before they can comprehend anything.

See problems 14-1, 14-2, and 14-3.

5. Summary

We have studied why $\partial T/\partial x$ in the thermally fully developed channel flows is a constant. The logical path that explains this fact can be more clearly depicted using a genetic analogy.

6. Reference

1. A. F. Mills, *Heat Transfer*, Prentice Hall, 1999.

7. Exercise Problems

14-1 Based on the logical exercise of genetics, and according to the following 1st-generation facts, prove that T(x) is a cubic-polynomial profile.
F1a: 1-D steady state heat conduction within a slab
F1b: $= q'''_{gen} = c_1 x$
F1c: Taylor's Series Expansion such as $T(i+1)=T(i) + (\Delta x) \dfrac{dT}{dx} + 0.5\,(\Delta x)^2 \dfrac{d^2T}{dx^2}$
F1d: First law of Thermodynamics

14-2 Consider the TFDF. Based on the logical exercise of genetics, and according to the following 1st-generation facts:
F1a: $\theta(\xi) = (T - T_s)/(T_m - T_s)$ is a function of ξ.
F1b: Ts is uniform

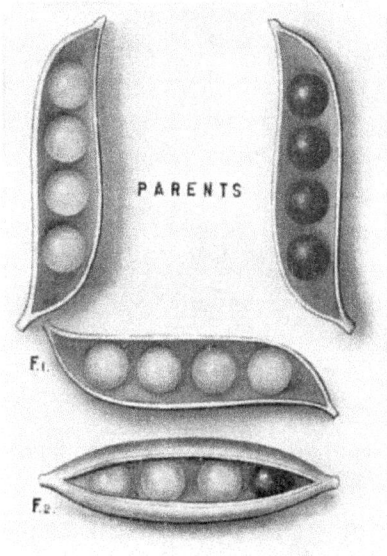

F1c: the energy equation, $u\dfrac{\partial T}{\partial x} = \alpha \dfrac{\partial^2 T}{\partial y^2}$,

prove that the energy equation can be transformed into:

$$\frac{d^2\theta}{d\xi^2} + Hu^*\theta = 0,$$

where $H = -Pe\,\dfrac{b}{\Delta T}\dfrac{dT_m}{dx} = $ constant, and $\Delta T = T_m - T_s$.

14-3 Which facts exactly lead to $dT_m/dx =$ constant for a TFDF with q"s = constant?

(a) is a function of ξ and q"s = constant,

(b) energy conservation over a segment-dx control volume, and q"s = constant,

(c) $\theta(\xi)$ is a function of ξ and energy conservation over a segment-dx control volume,

(d) $T_m - T_s = $ constant and q"s = constant.

Lessons 15

Internal Flows (III)—Thermal Aspect

I n this lesson, let us continue to study the thermal aspect in channel flows. Emphasis will be placed on derivations of temperature profiles, Nu, and some applications.

Nomenclature

b = half height of the channel

$c_1 = (h\Delta x/k)\,(p\Delta x/A_c)$

G = a dimensionless parameter, defined in Eq. (2), for TFDF and qs" = constant

H = a dimensionless parameter, defined in Eq. (14b), for TFDF and Ts = constant

Nu = h b /k, dimensionless h

Pe = Peclet number, $u_m b/\alpha$

pe = Peclet number, $u_m \Delta y/\alpha$

TFDF = acronym for thermally fully developed flow

$\Delta T = T_m - T_s$, C or K

uc = velocity at the centerline, m/s

$\theta = (T - T_s) / (T_m - T_s)$

---

1. Derivation of Nu Value for Uniform q''s

Having established Eq. (8) in Lesson 14 is a significant accomplishment for us. Now, we are pleasantly able to integrate Eq. (3) in Lesson 14 can be non-dimensionalized with respect to y with ease. For the sake of simplifying algebraic manipulations, Eq. (3) can be non-dimensionalized into

$$G(\xi^2 - 1) = \frac{d^2\theta}{d\xi^2},$$

(1)

$$\text{where} \quad G = -\left(\frac{u_c b}{\alpha}\right)\left(\frac{b}{\Delta T}\right)\frac{dT}{dx},$$

(2)

which is a dimensionless constant. Note that, in Eq. (2), we temporarily use the centerline velocity, not the mean velocity. Also,

$$\Delta T = T_m - T_s, \quad \theta = \frac{T - T_s}{\Delta T}, \text{ and } \xi = y/b.$$

Incidentally, G is positive all the time. For example, if the heating of the fluid takes place, ΔT is negative and dT/dx is positive. So, G is positive. Equation (1) can be integrated once and twice, with the aid of two boundary conditions, namely,

at $\xi = 0, \dfrac{d\theta}{d\xi} = 0$, and at $\xi = 1, \theta = 0$,

to eventually yield

$$\theta = G\left(\frac{\xi^4}{12} - \frac{\xi^2}{2} + \frac{5}{12}\right). \tag{3}$$

At this moment, G remains as an unknown. We can, however, use the definition of T_m to determine its value, of which the procedure will be left as a patient exercise. The final result is

$$G = \frac{1}{1.5D_1} = 3.0878, \tag{3a}$$

where $D1 = \int_0^1 (1-\xi^2)\left(\frac{\xi^4}{12} - \frac{\xi^2}{2} + \frac{5}{12}\right) d\xi = 0.2159,$

which can be either evaluated analytically or computed numerically. Finally, Eq. (3) can be written as

$$\theta = 3.0878\left(\frac{\xi^4}{12} - \frac{\xi^2}{2} + \frac{5}{12}\right). \tag{4}$$

Therefore, we have accomplished finding the temperature distribution T(x, y). Again, knowing T(x, y) is analogous to staying at home. At home, we get to leisurely open the refrigerator door, and lie on the couch. Having obtained T(x, y), we can find heat flux at the plate, h, and Nu.

$$h = q_S''/(T_m - T_s) = -\frac{k}{b}\left(\frac{d\theta}{d\xi}\right)_{\xi=1},$$

or

$$Nu = \frac{hb}{k} = -\left(\frac{d\theta}{d\xi}\right)_{\xi=1} = 2.0586, \tag{5}$$

in good agreement with the literature value, 2.0575 [1]. Incidentally, Eq. (5) can be obtained from the combination of Eq. (7) in Lesson 14 and Eq. (3a), too, without having to find the temperature profile given in Eq. (4). This remark is made to alert readers not to treat Eq. (3a) as an independent equation.

See Problem 15-1.

From the beginning of the problem all the way to the end, we have found h or Nu ourselves without having to rely on applied mathematicians or software programmers. Within the subject of heat convection, such an opportunity is quite rare.

2. Important Implications of Eq. (5)

As in the hydrodynamic consideration, we have obtained at least four important results:

(a) the temperature distribution, Eq. (4)

(b) Tc – Ts = 1.2866*$(T_m$ – Ts), which can be used to validate numerical schemes and experimental procedures. Tc stands for the temperature at the centerline.

(c) Based on Eq. (3a), if ΔT is given, then $\dfrac{dTm}{dx} = -\left(\dfrac{2.0586}{Pe}\right)\left(\dfrac{\Delta T}{b}\right)$, where $Pe = u_m b/\alpha$.

(d) $Nu = h\, b\, /\, k = 2.0586$ is a pure number, and is not a function of anything else.

(e) It is worth noting that we have two independent equations, Eq. (7) in Lesson 14, and Eq. (5) in this lesson, from which we can determine two unknowns. All other properties, conditions, and parameters must be given. Keeping this note in mind, we will be able to judge confidently if a given problem is under-specified or over-specified. No more, no less. If we cannot judge it quickly, we are out of the game.

Example 15-1

Given: thermally fully developed flow (TFDF) of water, incoming flow velocity is uniform and is at 0.04 m/s. The incoming mean temperature is 20C, and the plate temperature at the inlet is 90C. The heating flux on the plate into the fluid flow (in W/m^2) is maintained uniformly, but its value is not known.

Find: the total heating rate (in kW) and the average Ts (in C)

```
clc; clear
% heating of the water flow and qs=constant
b=0.01; Lx=5; Z=1; % geometry of the channel
k=0.61; rho=1000; cp=4180; aLf=k/(rho*cp);nu=9.5e-7;%fluid properties
um=0.04; Tm_in=20; Ts_in=90; % flow conditions
%
%>>>>>>>>>> data above are given
%
pr=nu/aLf; re=um*b/nu; %=421.05, physical parameters
Pe=pr*re; Nu=2.0586;
% calculation of the problem
Qs = Nu*k*Lx*(Tm_in - Ts_in)/b % = -43.95 kW
mflow = rho*b*Z*um;
Tm_ex= Tm_in - Qs/(mflow*cp)% =46.2865C
Ts_ex= Tm_ex + Ts_in - Tm_in % = 116.2865C
Ts_avg = .5*(Ts_in + Ts_ex) %= 103.14C
```

See also Problem 15-2.

---

3. Derivation of Nu value for uniform Ts

In heat-convection analyses, there may exist more problems subject to constant Ts than those subject to constant . Why did we present the latter first? There is a reason associated with such an arrangement. The reason lies in that it is more difficult to solve the case of constant Ts. Let us elaborate this reason below.

Equation (3) in Lesson 14 remains to be our starting point. It is not clear to us at this moment if the condition of TFDF is merely an assumption made by scholars, or if it is bound to happen in real life in the laboratory when the flow remains laminar and Ts is maintained constant. At least it is not clear to the author.

3-1 Justification

To proceed, however, let us assume that the TFDF condition exists. Hence, Eq. (2a) becomes

$$\frac{dT_m}{dx}\,\theta = \frac{\partial T}{\partial x} \text{ since } \frac{\partial T_S}{\partial x} = 0.$$

The equation above is equivalent to the first step of Separation of Variables, where we separate T(x, y) into a product of $T_m(x)$ and $\theta(y)$, as shown in Fig. 15-1.

Then, Eq. (3) in Lesson 14 temporarily becomes

$$u\,\theta\,\frac{dT_m}{dx} = \alpha\,\frac{\partial^2 T}{\partial y^2}. \tag{6}$$

Meanwhile, since we have already learned that

$$\frac{\partial^2 T}{\partial y^2} = \frac{\Delta T}{b^2}\frac{d^2\theta}{d\xi^2} \text{ , where } \Delta T = T_m - T_s,$$

Eq. (6) can be converted into

$$-\left[-Pe\,\frac{b}{\Delta T}\frac{dT_m}{dx}\right] = \frac{1}{\theta u^*}\frac{d^2\theta}{d\xi^2}. \tag{6a}$$

It is noticed that the right-hand side of Eq. (6a) is a function of ξ only, whereas the left-hand side is a function of x only. This fact leaves us no other options but claiming that both sides of Eq. (6a) must be equal to a constant. Thus, introducing

$$H = -Pe\,\frac{b}{\Delta T}\frac{dT_m}{dx} > 0, \tag{6b}$$

we obtain

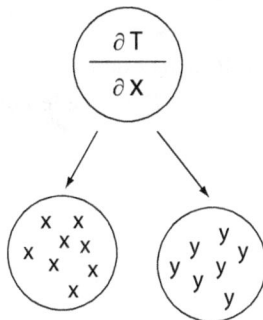

Fig 15-1   Separation of variables

$$\frac{d^2\theta}{d\xi^2} + Hu^*\theta = 0,$$

(7)

subject to: at $\xi = 0$, $d\theta/d\xi = 0$; at $\xi = 1$, Also, recall that $u^* = 1.5\,(1 - \xi^2)$.

3-2 Solution Procedure

Equation (7) is a second-order linear ordinary differential equation, and may look as if it can be readily handled by using a standard numerical method for solving two-point boundary-value problems. In fact, $\theta = 0$ for all ξ will be a trivial solution. We are, however, not interested in this trivial case.

To avoid obtaining a trivial solution, we may opt to give a guessed value for θ, like 1.3, at $\xi = 0$. We then proceed to solve Eq. (7), subject to the following two constraints:

(a) $1 = \int_0^1 u^*\theta\, d\xi$; (b) at $\xi = 0$, $d\theta/d\xi = 0$, by varying values of H and $\theta(0)$.

This algorithm may need to be involved with tremendous amount of numerical work, and is not too desirable. Having this concern, let us assume instead that

$$\theta = d_0 + d_1\xi + d_2\xi^2 + d_3\xi^3 + d_4\xi^4$$

(8)

Luckily, we can immediately drop d1, because of the condition: at $\xi = 0$, $d\theta/d\xi = 0$. Therefore, there are five unknowns: d0, d2, d3, d4, and H. We can establish five algebraic equations from the following conditions:

(a) at $\xi = 1$, $\theta = 0$;
(b) The definition of T_m that leads to $1 = \int_0^1 u^*\theta\, d\xi$;
(c) Integrating Eq. (7) once yields

$$\left(\frac{d\theta}{d\xi}\right)_{\xi=1} - \left(\frac{d\theta}{d\xi}\right)_{\xi=0} + H\int_0^1 u^*\theta\, d\xi = 0.$$

(9)

Since the second term is equal to zero due to line of symmetry, and the integral is equal to unity based on the definition of T_m (to be left as an exercise), Eq. (9) becomes

$$\left(\frac{d\theta}{d\xi}\right)_{\xi=1} + H = 0.$$

(9a)

(d) at $\xi = 0.5$, Eq. (7) must be satisfied.
(e) at $\xi = 0.9$, Eq. (7) must be satisfied.

Incidentally, locations chosen in conditions (d) and (e) are arbitrary. And such a treatment is known as the collocation method. Also, we should be alerted that Eqs. (d) and (e) are nonlinear equations.

The computation details are written in the Matlab code attached in the appendix. Only the final result is given here:

Nu = 1.8968 (in comparison with the literature value of 1.8850.)

It so happens that also H = 1.8968. The θ profile is plotted in Fig. 15-2.

A table including the internal flows in circular tubes lists Nu for two different cases for convenience.

	$q_s'' = $ const	Ts = const
Nu_b (plates)	2.0586	1.8968
Nu_D (tubes)	4.36	3.66

What if the boundary condition on the plate is neither $q_s'' = $ constant, nor Ts = constant? We may not be able to solve such problems analytically any more. Neither will Nu be a pure numerical number.

See Problem 15-3.

Example 15-2

Given: thermally fully developed water flow, incoming flow velocity is uniform and is at 0.04 m/s. The incoming mean temperature is 20C, and the plate temperature is maintained at 103.14C uniformly.

Find: the total heat rate (in kW) and Tm (in C) at the exit
Sol:

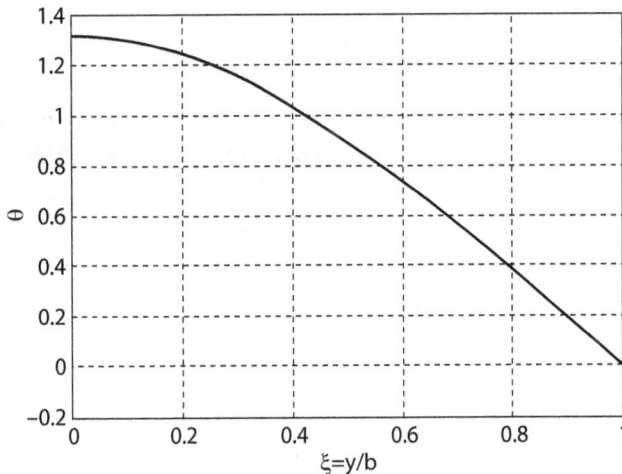

Fig. 15-2   Dimensionless temperature profile

```
%%
clc; clear
% heating of the water flow and Ts=constant
b=0.01; Lx=5; Z=1; % geometry of the channel
k=0.61; rho=1000; cp=4180; aLf=k/(rho*cp);nu=9.5e-7;%fluid properties
um=0.04; Tm_in=20; Ts=103.14; % flow conditions
pr=nu/aLf; re=um*b/nu %=421.05, physical parameters
nu=9.5e-7; pr=nu/aLf %=6.51
pe=pr*re; Nu=1.8968;
H=Nu; h=Nu*k/b % = 115.7
nx=100; nxp=nx+1; dx=Lx/nx;
% start calculation of the problem
for i=1:nxp;
 x(i)=(i-1)*dx;
 Tm(i)=Ts+(Tm_in-Ts)*exp(-H*x(i)/(pe*b));
end
sum=0;
for i=2:nxp
Tavg=.5*(Tm(i-1)+Tm(i)); qs=h*(Tavg-Ts)*dx*Z;
 sum = sum + qs;
end
Tm_ex=Tm(nxp)% =44.3178C
Qs = sum % = -40.659 kW
mflow = rho*b*um;
Qs_mass = mflow*cp*(Tm_ex-Tm_in) % = -40.659 kW
```

4. Let the Faucet Drip Slowly

Let us now study a water flow in which the velocity is so low that the heat conduction in the x direction can no longer be ignored in comparison with heat convection. Sometimes such a flow, shown in Fig. 15-3 is called slug flow. At the same time, there is a heat sink or a heat source in contact with the pipe wall. Let us further assume that T is not a function of r, but a function of x only.

We are often advised that in the winter, the faucet should be left dripping slowly to avoid frozen and burst pipes. It happens that this faucet-dripping phenomenon is quite analogous to the problem described in the paragraph above. The key issue is for us to see what a significant change of water temperature may occur if u increases from zero to a certain speed, even if this speed is extremely small. Then "what" is followed by "why".

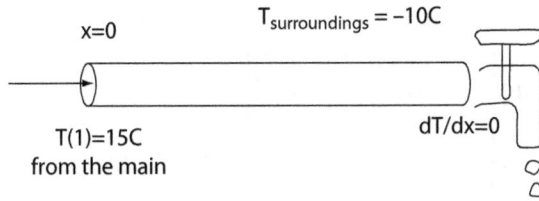

Fig. 15-3   System schematic of a dripping faucet connected with a slug pipe flow

The modeling of this flow problem is almost like that of a fin, but let us instead imagine that the material inside the fin becomes fluid-like and can move. In reference to Fig. 15-4,

$$q1 = \dot{m}\, c_p\, T_{i-1/2}, q2 = \dot{m}\, c_p\, T_{1+1/2}\,,$$

$$q_3 = kA_c\,(T_{i-1} - T_i)/\Delta x,\, q_4 = kA_c\,(T_i - T_{i+1})/\Delta x,$$

and $q5 = p\pi \Delta x h\,(T_i - T_{surr})$.

According to Fig. 15-4 that shows the energy balance over the control volume Δx, we can write:

$$q1 + q3 - q2 - q4 - q5 = 0,$$

leading to

$$0.5\, pe\,(T_{i-1} - T_{i+1}) + T_{i-1} - 2T_i + T_{i+1} - c_1\,(T_i - T_{surr}) = 0,$$

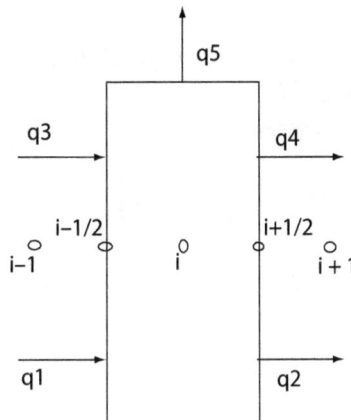

Fig. 15-4   Energy balance over the control volume, Δx

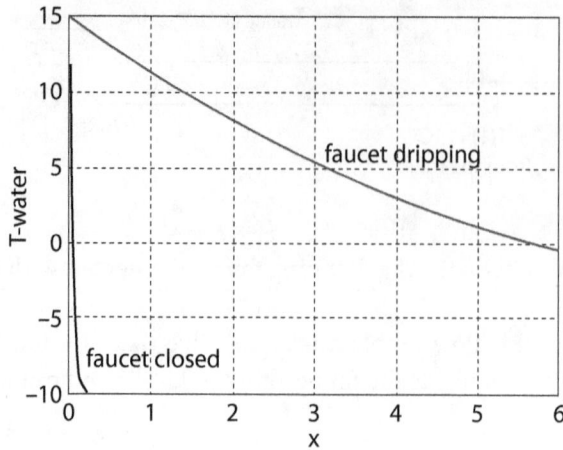

Fig. 15-5   Temperature of the water flow versus x.

which can be programmed and computed in a Matlab code given below. The computational result, T(x), is shown in Fig. 15-5.

```
% water dripping from the faucet in severely cold winter
clc; clear
% water properties
k=0.61; rho=1000; cp=4180; aLf=k/(rho*cp);
% geometries of the pipe
d=.017; p=pi*d; Ac=pi*d*d/4; h=2;
Lx=6; nx=40; dx=Lx/nx; nxp=nx+1;
x=linspace(0, Lx, nxp);
% parameters in the governing equation
bi=h*dx/k; ratio=p*dx/Ac; c1=bi*ratio;
for iu=1:2
% when iu=1, we examine the case of turning off the faucet completely.
um=(iu-1)*7e-4; pe=um*dx/aLf;
a(nxp,nxp)=0; b(nxp)=0;
T(1)=15; Tsurr=-10;
a(1,1)=1; b(1)=T(1);
for i=2:nx
a(i,i-1)=0.5*pe+1; a(i,i)=-2-c1;
a(i,i+1)=1-0.5*pe; b(i)=-c1*Tsurr;
end
a(nxp,nx)=-1; a(nxp,nxp)=1;
T=a\b';
plot(x,T); hold on; grid on;
```

xlabel('x'); ylabel('T-water')
end
text(3.2,5.7,'faucet dripping');
text(0.2, -7.5, 'faucet closed')
hold off

See Problem 15-4. In this problem, we will examine why such a small speed of water flow can avoid the disastrous pipe burst.

---

5. Summary

In this lesson, we have accomplished:

(a) analytical derivation of the Nu value in TFDF, subject to constant wall heat flux,
(b) numerical computation of the Nu value in TFDF, subject to constant wall temperature,
(c) investigation of faucet dripping phenomena in the winter.

---

6. References

1. Frank P. Incropera, David P. DeWitt, Theodore L. Bergman, and Adrienne S. Lavine, *Fundamentals of Heat and Mass Transfer*, John Wiley and Sons, 6th edition, 2006.

---

7. Exercise Problems

15-1 Find Nu for slug flows (u = constant) and the condition qs"=constant in TFDF regime between two parallel plates.

15-2 Given: water flow, TFDF, qs"=constant (but its value may not be known), k=0.61; ρ = 1000; cp = 4180; Z = 1;
Find: the exact number of variables that we must be given in order to solve the problem. These variables include inlet conditions, exit conditions, and channel geometries.

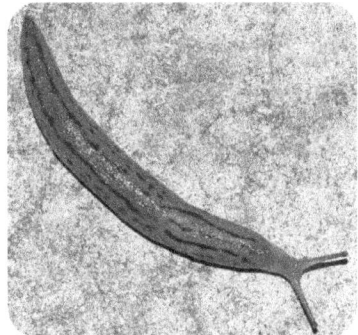

Note that, obviously, the problem cannot be under-specified. On the other hand, even though over-specifying allows us to solve the problem, it does not appear too professional.

15-3 We are given a problem in which neither q"s nor Ts is constant the TFDF regime. Determine what parameters Nu should depend on.

15-4 Revisit the faucet problem. We have computed T(x), and observed that, with a small speed of water flow (0.7 mm/sec, which is indeed slower than a slug's movement), we seem to be able to avoid water freezing. Exercise our brain, and identify our own reasons to explain the phenomenon.

Criteria for good reasons include:

(a) Not going beyond the scope of heat transfer such as using quantum physics.

(b) They do not need to be very complicated or laborious.

(c) They are involved with some computations and numbers.

8. Appendix

```
clc; clear % computation of the case of constant Ts
for i=1:5; q(i)=1; end % initial guesses
% dn = q1; d2=q2; d3=q3; d4=q4; H =q5; 5 unknowns, 5 eqns
% only dn*H is linearized. Luckily, the solution converged.
dnb=q(1); Hb=q(5);
% dnb = previously iterated value of dn
% Hb = previously iterated value of H
for iter=1:10
a41=1.125*Hb; a42=2+0.2813*Hb; a43=3+0.1406*Hb;
 a44=3+0.0703*Hb; a45=1.125*dnb;
%
a51=0.285*Hb; a52=2+0.2309*Hb; a53=5.4+0.2078*Hb;
 a54=9.72+0.1870*Hb; a55=0.285*dnb;
%
a=[1 1 1 1 0;... % theta = 0 at ksi = 1
 1-1/3 1/3-1/5 1/4-1/6 1/5-1/7 0; % definition of Tm
 0 2 3 4 1; % integrating the governing eqn
 a41 a42 a43 a44 a45; % collocate at ksi=0.5
 a51 a52 a53 a54 a55]; % collocate at ksi=0.9
%
b=[0 1/1.5 0 1.125*Hb*dnb 0.285*Hb*dnb];
q=a\b'; dnb=q(1); Hb=q(5);
```

```
q'
end
Nu= -(2*q(2)+3*q(3)+4*q(4)) % = 1.8968
Nu_literature = 7.54/4 %= 1.8850 (Nu_D = 7.54)
Hb %= 1.8968
dk=1/50;
for j=1:51
 ksi(j)=(j-1)*dk; y=ksi(j);
 theta(j)= q(1)+q(2)*y^2+q(3)*y^3+q(4)*y^4;
end
plot(ksi, theta); xlabel('\xi=y/b'); ylabel('\theta'); grid on
```

Lesson 16

Free Convection

In this lesson, let us turn our attention to free convection, a phenomenon that is somewhat different from forced convection, which was described in previous lessons.

Nomenclature

G_1 = a function of Pr, defined in Eq. (7)
Gr = Grashof number, defined in Eq. (3)
H = height of a wood block, m
$m_1 = \mu / (\Delta x)^2$
qs = supplied energy to keep the vertical plate at constant T
β = thermal expansion coefficient, 1/K
ρ = density, kg/m^3

1. Definition of Free Convection

In general, when (1) flows are subject to the gravitational force, (2) the density of the fluid varies due to a non-uniform temperature distribution, and (3) external agents such as fans, blowers, or pumps are absent, then free convection will take place. Perhaps the name "free" was derived from the fact that such flows are free of external agents. The term is interchangeable with "natural convection."

Before the refrigerators were invented, people used iceboxes that consisted of a well-insulated container, in which a block of ice was placed on the top shelf, shown in Fig. 16-1. The dense air close to the ice would sink and recirculate, chilling the food inside the container.

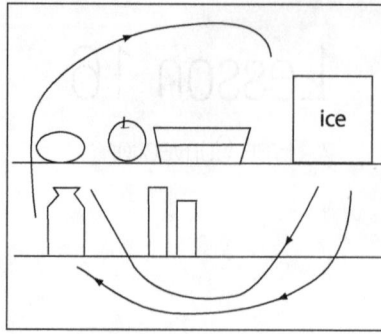

Fig. 16-1   The icebox in the pre-refrigerator era

2. Definition of Buoyancy Force

2-1 Buoyancy Force on an Object

To the best of our knowledge, primitive forces are body force, pressure force, shear-stress force, and normal-stress force. If the control volume contains a wood block and the water is quiescent, for example, then only the body force and the pressure force are present.

Consider a wood block immersed in water, shown in Fig. 16-2. We need to use our hand to exert a force, F, to hold the block still. Force balance over the block can be shown to be:

$$p1^*A - p2^*A - m^*g - F = 0. \tag{1}$$

It is worth noting that at this moment there is not such a force, termed "buoyancy force" appearing in Eq. (1). In other words, if the wood block is held perfectly still, there are exactly four force components acting on it, as shown. And only $p1^*A$ is acting upward to balance the other three downward forces.

Since p1 and p3 are at the same water level, and there is no flow moving between the two positions, p1 = p3. Similarly, p2 = p4. From fluid statics, we further know that

$$p1 - p2 = p3 - p4 = \rho_{water} gH.$$

Substituting the expression above into Eq. (1), and realizing that m = $A^*H^* \rho_{wood}$, we obtain

$$F = V^*(\rho_{water} - \rho_{wood})g. \tag{2}$$

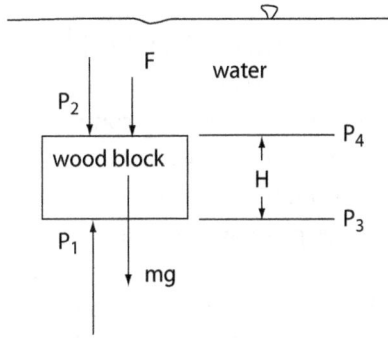

Fig. 16-2   Force balance over a wood block immersed in water

In some heat transfer textbooks, buoyancy force is defined as F. In Wikipedia, according to the article, Archimedes seemed to have defined buoyancy force as the weight of the fluid displaced by the object, which would have been $V^* \rho_{water} g$. We must be aware that these two definitions are not consistent.

When we speak of buoyancy force, we should not treat it as if it is an additional force mysteriously created from somewhere when a wood block is immersed in water or a balloon is floating in air. Instead, we should realize that if our hands are holding a balloon, there are only four forces acting on it, namely, F_hand, gravitational force, p_bottom, and p_top (if the air is windless).

2-2 Buoyancy Force on a Control Volume in the Flow

If the object is replaced with a small control volume dx*dy inside a flow, then there will be at least two modifications:

(a) We can no longer equate p1 – p2 with p3 – p4, as they may not be equal.
(b) There will be additionally shear stresses and normal stresses acting on the fluid inside the control volume.

Hence, the definition of buoyancy force becomes even more vague than in the case of an object immersed in a quiescent fluid, because we do not have an object, such as the emperor's gold crown, immersed in the fluid. Neither do we have Eq. (2) because of what we mentioned in modification (a) above.

If possible, perhaps the best way of technical communications is to simply use the phrase "a net uplifting force" instead of "buoyancy force."

When we chat with a junior high school student next time regarding an uprising airflow adjacent to a vertical hot plate, we may say, "....because the density of the air near the hot plate is smaller than that of the air far away, the air experiences a net uplifting force, driving it to rise." If we mention the term "buoyancy force," and if the student asks what buoyancy force is, we may be in an awkward position.

2-3 Density Variations

We must realize that it is density variations (in the flow field) that eventually drive the flow, not the existence of the gravitational force itself. Because density is not uniform in the flow field, each fluid chunk is thus subject to different magnitudes of gravitational forces. These gravitational forces, however, all act downward. How do some of them turn around to become uplifting forces?

Hey, are we here first or eggs?
Neither, worms are here first.

Mass conservation therefore enters the issue to play an important role. Because the mass must be conserved, different values of negative v's (pulled down by different values of gravitational forces) will eventually generate different values of u's. But not all fluid chunks can sink or spread horizontally forever. At a certain juncture, some chunks are bound to turn up to fill up the void. These bumping, squeezing, and expanding motions subsequently create non-uniform stresses and pressures, in addition to original non-uniform hydrostatic pressure. Chicken-and-egg effects thus ensue.

3. The Main Difference between Free Convection and Forced Convection

In free convection problems, the gravitational force exists. In force convection problems, the gravitational force is neglected. Therefore, based on the argument above, we state that the existence of $\rho(x,y)*g*dx*dy$ makes the main difference between two types of flows.

The density, $\rho(x,y)$ is intimately related to $T(x,y)$ in fluids. Therefore, the coupling between the momentum equation and the energy equation appears.

In free convection, we must solve $u(x,y)$, $v(x,y)$, $p(x,y)$, and $T(x,y)$ simultaneously. In forced convection, often we are able to solve u, v, and p first without the knowledge of T.

Another minor difference is that, in forced convection, *Re* number is a key parameter; whereas in free convection, *Gr* number is a key parameter, which will be presented next.

4. How Does Gr Number Arise?

Let us select two representative terms, the momentum rate change and the body force, in the y-direction governing equation.

$$\rho\Delta v^2 \approx -\rho g\Delta y,$$

which can be modified with a trivial manipulation as

$$\rho\Delta v^2 \approx -\rho g\,\Delta y + \rho_0 g\Delta y - \rho_0 g\Delta y$$

or

$$\rho\Delta v^2 \approx (\rho_0 - \rho)g\,\Delta y - \rho_0 g\Delta y.$$

We then divide both sides by ρv_0^2 to yield

$$\Delta v^{*2} \approx [(\rho_0/\rho - 1)(g\Delta y/v_0^2)] - (\rho_0/\rho)\,(g\Delta y/v_0^2)$$

Since our current interest is on the appearance of *Gr*, let us only focus on the term in the bracket.

In most incompressible flows, it is safe to accept $\rho_0 T_0 = \rho T$ for ideal gases. Furthermore, it is our prerogative to choose the magnitude of v_0. If we choose v_0 to be v/x_0, we can force the Reynolds number to be unity, hence eliminating its presence in the governing equation. Within the frame of these thoughts, we obtain

$$[(\rho_0/\rho - 1)\,(g\Delta y/v_0^2)] = \left(\frac{T - T_0}{T_0}\right)\left(\frac{gx_0^3}{v^2}\right)\Delta y^*.$$

It can be proved that, for ideal gases, the thermal expansion coefficient, defined as

$$\beta = \frac{1}{\tilde{v}}\left(\frac{\partial\tilde{v}}{\partial T}\right)_p \approx \frac{1}{T_0},$$

where $\tilde{v}$ is specific volume (v is reserved for the y-direction flow velocity). Thus, we finally obtain

$[(\rho_0/\rho-1)\,(g\Delta y/v_0^2)] = \theta\,Gr\,\Delta y^*$, where

$\theta = (T - T_0)/\Delta T$ and $Gr = \dfrac{g\beta\Delta Tx_0^3}{v^2}$. $\hspace{3cm}$ (3)

Incidentally, the Raleigh number, Ra, is defined as Gr^*Pr, and is sometimes used in replacement of Gr. Approximately, Gr_critical = 10^9, beyond which the flow will become turbulent.

Example 16-1

What is a typical value of Gr for a typical free convection airflow over a flat plate if $\Delta T = T_s - T_\infty = 30$?

Sol:
clc; clear
g=9.8; Tn=300; beta=1/300;
dT=30; xn=0.5;
new=1.6e-5;
Gr=(g*beta*dT*xn^3)/new^2 % = 4.785e8, most likely laminar

5. δ_T and δ in Free Convection

In forced convection, δ_T and δ can compete with each other in terms of their thicknesses, depending on the values of the Prandtl number, Pr. For example, δ_T for liquid metals can be much greater than δ. Reversely, δ for motor oils can be much thicker than δ_T.

In free convection, the situation is different. See Fig. 16-3 for reference. While it is still true that δ for motor oils can be much thicker than δ_T, it is no longer true that δ_T for liquid metals can be much greater than δ. The reason is that, in free-convection boundary-layer flows, there will be always fluid movements, however small, as long as there exist temperature gradients. Let us try to remember:

There is a 7-year-old-children birthday party; there is plenty of noise.
There is a will; there is a way.
There are temperature variations; there are fluid movements.

In short, δ cannot possibly be thinner than δ_T. Also see the sketches above. If we find it difficult to remember the rationale of this phenomenon, try to identify the subscript "T" with "thin." Namely, δ_T is the thin guy.

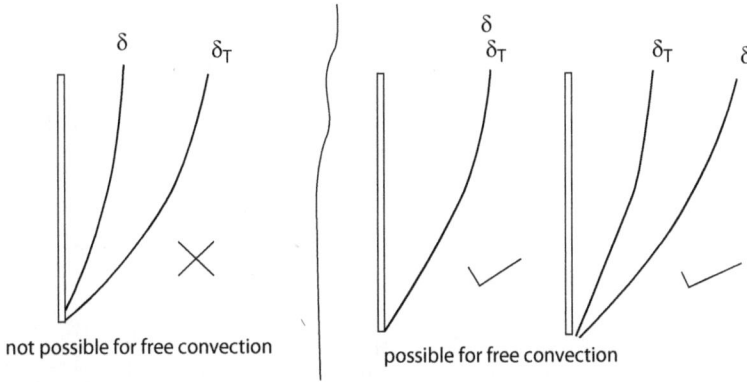

Fig. 16-3   Possible and not possible situations for free convection

6. A Four-Cell Buoyancy-Driven Flow in a Square Enclosure

In this section, let us examine a free convection problem of the airflow recirculating inside a 2-D square enclosure. Most of the terms will be dimensional, so that we can quickly get into the computation, and readily compare the magnitudes of various terms. If we have to rely on Bob, the applied mathematician, it may be advisable for us to Ndm the governing equations first.

6-1 Description

In reference to Fig. 16-4, the left face of the enclosure is hot; the right face is cold. The airflow will recirculate clockwise. There are only 4 computational cells. Hence the accuracy of the solution is not expected. But the qualitative trend can be observed. All the relevant data are given in the Matlab code in the Appendix.

It is our choice to locate T and ρ at the node of v. There are totally 16 unknowns and 16 governing equations. They are

u1, u2, ua, v1, v2, va, p1, p2, p3, p4, T1, T2, Ta, ρ1, ρ2, and ρa.

Values at node *a* are simply averages of the two neighboring nodal unknowns. For example, Ta = 0.5*(T1 + T2). The convection terms are linearized using Taylor's Series Expansion. A matrix of size 16 is identified, and the system of equations is solved by a direct solver. Hence the method is equivalent to the Newton-Raphson method. The solution has converged within ten iterations.

For simplicity, $\Delta x = \Delta y$ is taken. Also, *p*1 is used as the reference pressure, which is arbitrarily taken as 2. The temperatures and densities are located at the same nodal positions as v's according to the author's personal preference.

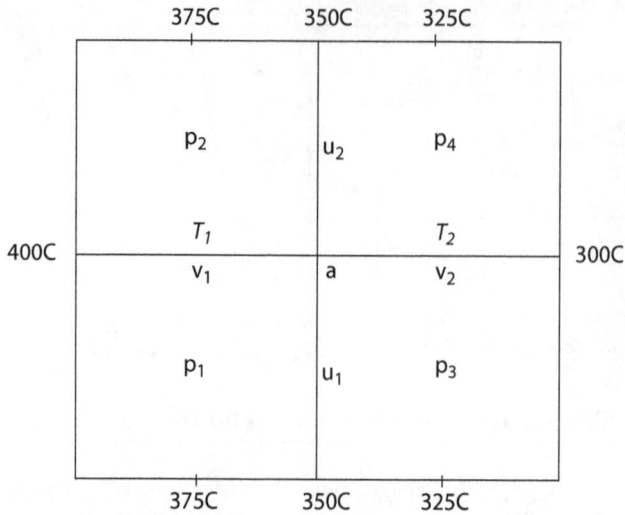

Fig. 16-4 **A four-cell computational grid system representing a free-convection flow inside a 2-d enclosure**

6-2 Derivation of Governing Equations

For the governing equation of v1, we can write:

convection = viscous stress – pressure gradient – body force,

where

convection = $\overline{\rho}_1 v_1 (v_1 - 0)/\Delta y$, where "overbar" stands for "previously iterated values"
stresses = $\mu(0 - 2^*v1 + va) / (0.5\,\Delta x)^2 + \mu(0 - 2^*v1 + 0) / (\Delta y)^2$
pressure gradient = $(-p1 + p2) / \Delta y$,
body force = $\rho_1{}^*g$.

With $m_1 = \mu/(\Delta x)^2$ and $v_1^2 \gg -\overline{v}_1^2 + 2\,\overline{v}_1 v_1$ linearized by Taylor's Series Expansion, we can derive:

$$\left(-10m_1 - \frac{2\overline{v}_1\overline{\rho}_1}{\Delta y}\right)v_1 + (4m_1)\,v_a + \frac{p_1}{\Delta y} - \frac{p_2}{\Delta y} - g\rho_1 = -\overline{\rho}_1(\overline{v}_1^2/\Delta y). \tag{4}$$

It seems safe not to include ρ_1 in the linearization, because ρ does not vary too much. Other governing equations can be derived similarly.

See Problem 16-1.

6-3 Discussions of the Result

(a) The result looks qualitatively reasonable.
(b) The code is validated to make sure that it is bug-free.
   See Problem 16-2.
(c) Increase $T_{hot} - T_{cold}$ to see if increases.
(d) If there is no flow, what values of p1, p2, p3, and p4 do we expect?
   See Problem 16-3.

7. Does Lighting a Fire in Fireplace Gain Net Energy for the House?

Lighting a fire in the fireplace may be cozy or romantic for family gatherings and a couple's date. Some people think that that is about it; they have doubts in their minds that it can actually gain net energy for the house.

In this section, let us conduct a simple free-convection analysis to see if their doubt is valid.

One of the assumptions is that the mass flow rate is steady. All the hot air escaping from the chimney is replaced by fresh outdoor cold air that manages to seep through various gaps of the house. In fact, this assumption is probably a good one, because otherwise very soon the house would become a highly depressurized space. See Fig. 16-5 for reference.

All the relevant data are given in the Matlab code.

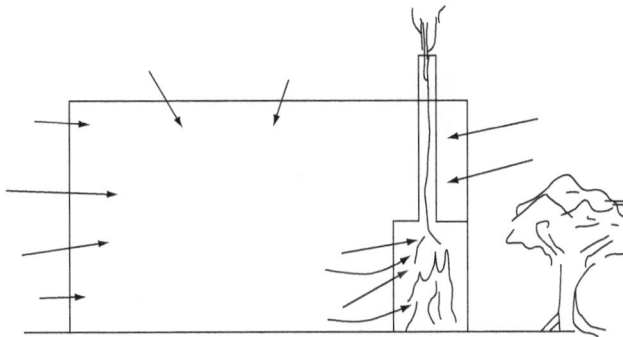

Fig. 16-5   The hot airflow exiting the chimney is assumed to be equal to the cold airflows seeping through gaps of the house.

7-1 Governing Equations

Let us take the house as the control volume, and treat the problem as a steady state one. Thus,

$$\dot{m}_{in} = \dot{m}_{exit} = \rho_c A_c v_{exit}.$$

The subscript "c" stands for "chimney."
The first law of thermodynamics dictates

$$m_{in} C_p T_\infty - m_{exit} C_p T_{exit} - \dot{Q}_{combustion} = \left(\frac{dU}{dt}\right)_{house}. \tag{5}$$

If $dU/dt > 0$, the house is gaining energy. If $dU/dt < 0$, the house is losing energy.

First, let us estimate the exit airflow velocity from the chimney. The difference between two pressures separated by a vertical air column of height H is:

$$\Delta p = \rho_0 \, g \, H,$$

which is equal to the pressure drop driving the momentum rate difference between the entrance and the exit.

Finally, a burning rate of wood logs at 5 kg/hr in the fireplace is assumed.

7-2 Nomenclature in the Code

H: height of the chimney, m

mflow: mass flow rate (= sum of all the cold air seeping into the house = m_exit from chimney)

new: kinematic viscosity of air

rhoc: average density inside the chimney

rhon: density of air at 300K

Ac: the cross-sectional area of the chimney (with a half width b and depth H)

Tinf: outdoor air temperature, 273K

7-3 Matlab Code

```
clc; clear
g=9.8;
new= 4.56e-5; % at T=550K
cp = 1005;
rhon=1.1614; % at T=300K
c1=300*rhon;
```

rhoc=c1/550 % = 0.6335 the average density of air inside the chimney
b=0.1; Z=.5; H=5.272; Ac=Z*2*b;
% estimate the frictional force
um=10; Tex=500; Tinf=273;
re=um*b/new
tauw = 6*0.5*rhoc*um*um/re
friction=tauw*H*Z*2 % = 0.053
%
dp=rhon*g*H
vex = sqrt(dp/rhoc) % = 9.7324 m/s
mflow=rhoc*Ac*vex
Q_conv = mflow*cp*(Tex - Tinf) % = 1.41e5 Watt
% If Tex=306.6, Q_conv ~ Q_combustion
%
HC=1.5e7;
mburn=5; % kg/hr
Q_combustion = mburn*HC/3600 % = 2.08e4 Watt
%
% So, Q_conv > Q_combustion, implying that the house is losing energy.

8. Solar-Radiation-Ice Turbine

A daydreaming amateur has proposed an idea that uses the abundant supply of ice in the winter and the abundant supply of solar radiation in the summer. These two supplies are absolutely free and clean. The system schematic is shown in Fig. 16-6.

8-1 Description of the Machine

The ice blocks are stored in a well-insulated underground storage space throughout the year. The height of the enclosure can be approximately 20m.

We will focus on only the thermal aspect of the problem. No cost estimate is conducted.

The thermal efficiency of this machine cannot possibly exceed:

$$\eta_{th} = 1 - T_L/T_H = 1 - 273 / (273+70) \approx 20 \%,$$

due to the small temperature difference. For internal combustion engines, for example, the limit is $1 - 300/(273+2500) \approx 90\%$.

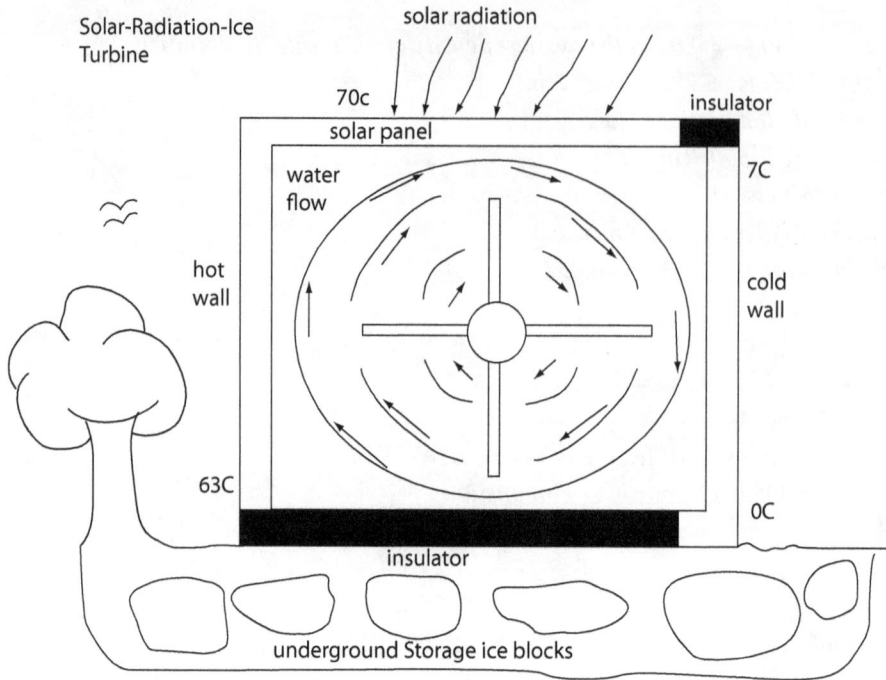

Fig. 16-6 **A proposed turbine driven by buoyancy flows, of which the temperature difference is caused by solar radiation and ice blocks.**

8-2 Some Analyses

A typical computational cell will be similar to v1 computational cell given in Section 6-1. We have already learned how to derive the algebraic governing equations. The remainder of the work seems to expand four cells into, perhaps, 1,000 cells. Skillful programming is therefore expected.

Questions that can be raised include:
1. What is a typical momentum rate for such a system?
2. Is this rate sufficient to drive the propeller (to overcome the friction of gear transmissions)?
3. If it is, how much Kw is the system able to generate?
4. If a small model in the laboratory is to be built, what parameters should be kept constant in order to ensure that the prototype is well predicted?

Similar to the correlation of Nu for forced convection over a horizontal flat plate, there are correlations available in the literature for free convection over a vertical plate. In [1], the following correlation was reported:

$$\overline{Nu} = 0.7071 \, Pr^{0.5} \, [Gr/G_1(Pr)]^{0.25}, \tag{7}$$

where $G_1 = 0.609 + 1.221 \, Pr^{0.5} + 1.238 \, Pr$.
See Problem 16-4.

Example 16-2

Consider an airflow moving over a vertical plate of L = 0.5m and Z=1m. The temperatures of the plate and the ambient air are known to be 57C and 27C, respectively. Find the heat transfer (in W) from the plate to the ambient air if Ts is kept constant. See Fig. 16-7 for reference.

```
clc; clear
g = 9.8;
k=0.026; new=1.6e-5; Pr=0.73;
Lx =0.5; Z=1;
Ts=57; Tinf=27; dT=Ts-Tinf; beta=1/(Tinf+273);
Gr = g*beta*Lx^3*dT/new^2 % = 4.7852e8
G1=0.609 + 1.221*Pr^0.5+1.238*Pr;
Nu = 0.7071*Pr^.5*(Gr/G1)^0.25 % = 70.669
h = k*Nu/Lx % = 3.6748
qs = h*dT*Lx*Z % = 55.122 W
```

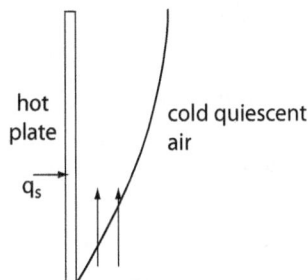

Fig. 16-7   Free convection over a vertical plate

10. Summary

In this lesson, we have learned

(a) how to take the force balance over a wood block immersed in water,
(b) to be cautious when mentioning the term "buoyancy force,"
(c) a four-computational-cell problem of recirculating flows, driven by buoyancy force, inside a 2-D enclosure,
(d) that lighting fires in the fireplace inside our houses may not necessarily gain energy for us.

11. Reference

1. E. J. LeFevre, Laminar free convection from a vertical plane surface, *Proc. Ninth Int. Congr. Appl. Mech.*, Brussels, vol. 4, 168, 1956.

12. Exercise Problems

16-1 Derive the governing equation for v_2. The result should be similar to Eq. (4).

16-2 Check residuals for the other nine algebraic governing equations to see if they are indeed zeros.

16-3 What is the value of the net force acting on the fluid within control volume, assuming Z=1m?

16-4 Prove that, for boundary-layer flows over a vertical plate maintained at constant temperature, Nu, is a function of *Gr* and *Pr* only.

13. Appendix

```
% Buoyancy-Driven Recirculating Flows in a 2-D Enclosure
clc; clear
icheck=1; % if icheck=1, check the correctness of six governing equations.
rhon=1.1614; g=9.8; new=1.6e-5; mu=new*rhon; k=0.026; cp=1005;
L=0.01; dx=L/2; dxs=dx*dx; dy=dx;
TL=400; TR=300; dT=(TL-TR)/4; p_ref=2;
Tf=TL-dT; Tg=Tf-dT; Th=Tg-dT;
c1=300*rhon; m1=mu/dxs; k1=k/dxs; rhog=c1/Tg;
a(16,16)=0; b(16)=0; % initializing matrix coefficients for convenience
```

```
% initial guess
% u1=q1; u2=q2; ua=q3; v1=q4; v2=q5; va=q6; p1=q7; p2=q8; p3=q9;
% p4=q10; T1=q11; T2=q12; Ta=q13; rho1=q14; rho2=q15; rhoa=q16;
u1b=0; u2b=0; uab=0; v1b=0; v2b=0; vab=0;
T1b=300; T2b=300; Tab=300; rho1b=rhon; rho2b=rhon; rhoab=rhon;
for iter=1:30
 ru1=0.5*(rhoab+rhog)/dx; ru2=ru1;
 convu1= ru1*(u1b*u1b);
 convu2=-ru2*(u2b*u2b);
 convv1=-rho1b*(v1b*v1b/dy);
 convv2= rho2b*(v2b*v2b/dy);
 rc1=rho1b*cp/dy; rc2=rho2b*cp/dy;
 convT1=-rc1*v1b*T1b;
 convT2= rc2*v2b*T2b;
a(1,1)=-10*m1+2*u1b*ru1; a(1,3)=4*m1; a(1,7)=1/dy; a(1,9)=-1/dy;
 b(1)=convu1;
a(2,2)=-10*m1-2*u2b*ru2; a(2,3)=4*m1; a(2,8)=1/dy; a(2,10)=-1/dy;
 b(2)=convu2;
a(3,1)=1; a(3,2)=1; a(3,3)=-2;
a(4,4)=-10*m1-2*v1b*rho1b/dy; a(4,6)=4*m1; a(4,7)=1/dy; a(4,8)=-1/dy;
 a(4,14)=-g; b(4)=convv1;
a(5,5)=-10*m1+2*v2b*rho2b/dy; a(5,6)=4*m1; a(5,9)=1/dy; a(5,10)=-1/dy;
 a(5,15)=-g; b(5)=convv2;
a(6,4)=1; a(6,5)=1; a(6,6)=-2;
a(7,1)=-1; a(7,4)=-1; a(7,7)=1; b(7)=p_ref; % let p1=1 as p_reference.
a(8,2)=-1; a(8,4)=1;
a(9,1)=1; a(9,5)=-1;
a(10,2)=1; a(10,5)=1;
a(11,4) =-rc1*(T1b-Tf); a(11,11)=-10*k1-rc1*v1b; a(11,13)=4*k1;
 b(11)=-2*k1*Tf-4*k1*TL+convT1;
a(12,5) = rc2*(T2b-Th); a(12,12)=-10*k1+rc2*v2b; a(12,13)=4*k1;
 b(12)=-2*k1*Th-4*k1*TR+convT2;
a(13,11)=1; a(13,12)=1; a(13,13)=-2;
a(14,11)=rho1b; a(14,14)=T1b; b(14)=c1+rho1b*T1b;
a(15,12)=rho2b; a(15,15)=T2b; b(15)=c1+rho2b*T2b;
a(16,14)=1; a(16,15)=1; a(16,16)=-2;
q=a\b';
% need to update
u1b=q(1); u2b=q(2); uab=q(3); v1b=q(4); v2b=q(5); vab=q(6);
T1b=q(11); T2b=q(12); Tab=q(13); rho1b=q(14); rho2b=q(15); rhoab=q(16);
```

```
q(1:10)' % monitor the convergence
end
q'
% check the correctness of the algebraic governing equations
resM=0;
if (icheck ==1)
% v1 equation
diffu=-m1*10*q(4)
bodyF=q(14)*g
dpdy=(q(7)-q(8))/dy
conv=q(14)*q(4)*q(4)/dy
res=diffu-bodyF+dpdy-conv; resM=max(resM, abs(res));
% v2 equation
diffu=-m1*10*q(5)
bodyF=q(15)*g
dpdy=(q(9)-q(10))/dy
conv=q(15)*q(5)*(-q(5))/dy
res=diffu-bodyF+dpdy-conv; resM=max(resM, abs(res));
% u1 equation
diffu=-m1*10*q(1)
rho1=.5*(q(16)+rhog);
dpdy=(q(7)-q(9))/dx
conv=rho1*q(1)*(-q(1))/dy
res =diffu+dpdy-conv; resM=max(resM, abs(res));
% u2 equation
diffu=-m1*10*q(2)
dpdy=(q(8)-q(10))/dy
conv=rho1*q(2)*q(2)/dy
res =diffu+dpdy-conv; resM=max(resM, abs(res));
% T1 equation
diffu=k1*(-10*q(11)+2*Tf+4*TL+4*Tab)
conv=q(14)*cp*q(4)*(-Tf+q(11))/dy
res=conv-diffu; resM=max(resM, abs(res));
% T2 equation
diffu=k1*(-10*q(12)+2*Th+4*TR+4*Tab)
conv=q(15)*cp*q(5)*(-q(12)+Th)/dy
residual=conv-diffu; resM=max(resM, abs(res));
end
resM
```

Lesson 17

Turbulent Heat Convection

As Reynolds numbers increase in boundary-layer flows or in channel flows, the regular patterns of flows will start becoming irregular. Most of the field variables in the flow, such as u, v, w, p, T, and ρ will exhibit fluctuations. Such flows are known to be turbulent flows.

To attempt to study turbulent flows and heat transfer well, scholars have devoted considerable effort to conducting modeling, numerical simulations, and experiments. For college education, however, it is sufficient for students to be aware of some fundamentals.

Nomenclature

B_2 = a constant, defined in Eq. (8) $-\dfrac{1}{\mu}\dfrac{dp}{dx}$,

C_1 = a constant, defined in Eq. (11) for circular tubes

k = turbulence kinetic energy, used in the k-ε model, J

L = van Driest mixing length, m

u_τ = a reference velocity widely used in turbulent-flow analyses, defined in Eq. (7b)

y^+ = a dimensionless y, defined in Eq. (7a)

α_t = turbulent thermal diffusivity, sometimes $\alpha_t \approx v_t$, m^2/s

ε = dissipation of turbulence kinetic energy, used in the k-ε model

ϕ = any instantaneous field variable

$\bar{\phi}$ = time-averaging value of ϕ

ϕ' = fluctuation of ϕ

v_t = turbulent viscosity, defined in Eq. (4b), m^2/s

1. Introduction

Let us be aware of some basic characteristics of turbulent flows. The following categorization gives us some ideas:

1-1 Speed of Typical Flows

0–20 m/s: laminar flows over flat plates (if L < 0.5m)
20–100 m/s (or Ma < 0.3): incompressible turbulent flows over flat plates
100–343 m/s: subsonic compressible turbulent flows

1-2 Frequencies of Turbulence and Molecular Collision

Turbulence: $1 - 10^4$ per sec
molecular collision: 5×10^9 per sec
Therefore, turbulent activities are much slower than those at the molecular level.

1-3 Superposition

In turbulent flows, a certain field variable, ϕ, representing u, v, w, p, T, or ρ, can be split into two parts as

$$\phi = \bar{\phi} + \phi',$$

where is the instantaneous variable, $\bar{\phi}$ is the time-averaging variable, and ϕ' is the fluctuation, as shown in Fig. 17-1. By definition, we have

$\bar{\phi} = \int_0^{t_0} \phi' \, dt = 0$, where t_0 is an appropriate time interval.

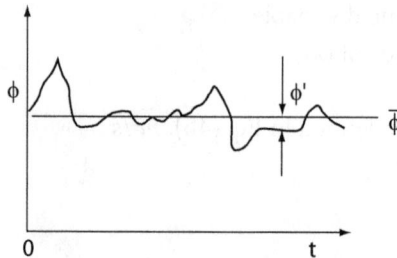

Fig. 17-1   Graphic representations of the three turbulent variables

2. A Fundamental Analysis

An analysis of turbulent flows requires us to establish governing equations first.

2-1 Governing Equations

Ignoring the density variations, let us derive the x-direction momentum equation in turbulent flows as follows:

$$\frac{\partial}{\partial y}(uv) \approx \nu \frac{\partial^2 u}{\partial y^2} - \frac{1}{\rho}\frac{\partial p}{\partial x}.$$ (1)

Here we assume that the variations of u or u^2 in the x direction are small.

Since $= \bar{u} + u'$, $v = \bar{v} + v'$, and $p = \bar{p} + p'$, we can change Eq. (1) by straightforward substitutions into

$$\frac{\partial}{\partial y}[(\bar{u}+u')(\bar{v}+v')] \approx \nu \frac{\partial^2(\bar{u}+u')}{\partial y^2} - \frac{1}{\rho}\frac{\partial(\bar{p}+p')}{\partial x}.$$ (2)

Taking time average over Eq. (2), and realizing that, for example,

$$\overline{\frac{\partial}{\partial y}(\bar{u}v')} = \frac{\partial}{\partial y}(\bar{u}\bar{v}') = \frac{\partial}{\partial y}(\bar{u}\,0) = 0,$$

we obtain

$$0 \approx \nu \frac{\partial^2 \bar{u}}{\partial y^2} - \frac{\partial}{\partial y}\overline{(u'v')} - \frac{1}{\rho}\frac{\partial \bar{p}}{\partial x}.$$ (3)

The reason that $\bar{u}'=0$ and $\bar{v}'=0$, but $\overline{u'v'}\neq 0$ can be illustrated by the following table:

t	0	1	2	3	4	5		
u'	1	2	3	-1	-2	-3	$\bar{u}'=0$	
v'	-2	-4	-3	3	3	3	$\bar{v}'=0$	
$u'v'$	-2	-8	-9	-3	-6	-9	$\overline{u'v'}$	= -6.16 $\neq 0$

In fact, not only is $\overline{u'v'}$ unequal to zero, but also usually its value is negative at least for cases where $\partial \bar{u}/\partial y > 0$. To justify $\overline{u'v'} < 0$, we can imagine that it is more probable for a fluid chunk to tend to preserve its momentum. At a certain instant, let us say that the time-average velocities, $\bar{u}$ and $\bar{v}$, in a steady state turbulent flow are 30 m/s and 1 m/s, respectively. Further assume that $|u'| = |v'| = 0.2$ m/s. Then, let us tabulate possible values of $u'v'$ as:

30.2 x 1.2 = 36.24 if $u' > 0$ and $v' > 0$... less probable
29.8 x 1.2 = 35.76 if $u' < 0$ and $v' < 0$
30.0 x 1.0 = 30.00 if $u' = v' = 0$
30.2 x 0.8 = 24.16 if $u' > 0$ and $v' < 0$
29.8 x 0.8 = 23.84 if $u' < 0$ and $v' > 0$... less probable

zero equation, one equation, two-equation models, Which one should I pick?

In light of this rationale, and according to experiments, we observe that near the wall where $\partial u/\partial y > 0$, the fluctuations are generally proportional to the time-average velocity gradient. Introduction of the Prandtl mixing length, L, yields

$|u'| = L\,\partial u/\partial y$, $|v'| = L\,\partial u/\partial y$, and

$$-\overline{u'v'} = L^2\,(\partial u/\partial y)^2 = v_t\,(\partial u/\partial y), \tag{4a}$$

where $v_t = L^2\,(\partial u/\partial y)$. $\tag{4b}$

The time-averaged overbar has been dropped for convenience. There is no need to carry the overbar around all the time. Hereafter, u is meant to be the time-averaged velocity.

With the aid of Eq. (4a), we can readily cast Eq. (3) into the following form:

$$-\frac{\partial}{\partial y}\left[(v + v_t)\frac{\partial u}{\partial y}\right] = -\frac{1}{\rho}\frac{\partial p}{\partial x}$$

or

$$-\frac{\partial}{\partial y}\left[\left(1 + \frac{vt}{v}\right)\frac{\partial u}{\partial y}\right] = -\frac{1}{\mu}\frac{\partial p}{\partial x} \tag{5}$$

2-2 Zero-Equation Turbulence Model

If $\overline{u'v'}$ can be expressed by existing field variables, then the formulation is called zero-equation turbulence models, as we do not need to derive any extra equations to express $\overline{u'v'}$.

A widely used model is known as k- model, in which $\overline{u'v'}$ is expressed in terms of k and ε, which in turn are governed by k transport equation and ε transport equation.

In this textbook, we will simply focus on the van Driest model [1], in which the Prandtl mixing length is expressed by

$$L = 0.4y\,[\,1 - \exp(-\,y^+/A)], \tag{6}$$

where $y^+ = y u_\tau / v$ and $u_\tau = (\tau_s / \rho)^{1/2}$

The parameter A is an empirical damping factor. Its value is 26. It can be varied slightly for various turbulent flow situations.

2-3 Discretized Governing Equation

The right-hand side of Eq. (5) should be equal to a positive constant, B_2, for fully developed turbulent channel flows. Thus, after taking a further differentiation, we have

$$-\left(1+\frac{v_t}{v}\right)\frac{d^2u}{dy^2} - \frac{du}{dy}\frac{d}{dy}\left(\frac{v_t}{v}\right) = B_2 \text{ , or after algebra,}$$

$$-u_{j-1} + 2u_j - u_{j+1} = B_3 (\Delta y)^2,$$

$$(8)$$

where $B_3 = (B_2 + E_1)/E_2$, $E_1 = \dfrac{du}{dy}\dfrac{d}{dy}\left(\dfrac{v_t}{v}\right)$, and $E_2 = 1 + \dfrac{v_t}{v}$

Based on Eq. (5), after Ndm, we know that B_3 is a function of Re. Furthermore, Eq. (8) is a highly nonlinear equation. If we use an iterative method to solve it, and if the solution will converge, then we are very lucky. It turns out that we are indeed very lucky. With some under-relaxation, we are able to render the solution converging, as shown in the Matlab given in the appendix.

In the thermal aspect, if we assume that, for turbulent flows near the wall, $\partial T/\partial x \approx 0$, then the energy equation reduces to

$$\frac{\partial}{\partial y}\left[(\alpha + \alpha_t)\frac{\partial T}{\partial y}\right] = 0,$$

$$(9)$$

or

$$-\left(\frac{1}{pr} + \frac{\alpha_t}{v}\right)\frac{d^2T}{dy^2} - \frac{dT}{dy}\frac{d}{dy}\left(\frac{\alpha_t}{v}\right) = 0,$$

$$\text{or } -T_{j-1} + 2T_j - T_{j+1} = B_4 (\Delta y)^2,$$

$$(10)$$

where $\quad B_4 = E_3/E_4$, $E_3 = \dfrac{dT}{dy}\dfrac{d}{dy}\left(\dfrac{\alpha_t}{v}\right) \quad$ and
$E_4 = \dfrac{1}{Pr} + \dfrac{\alpha_t}{v}$

If $\alpha_t \approx v_t$ can be assumed, Eq. (10) can be numerically solved similarly to Eq. (8). See Exercise Problem 17-1.

In order to learn how to solve Eq. (8), we need to know how the laminar counterpart is solved. For laminar flows, we have $v_t = 0$. So, $B_3 = B_2$.

In addition, we need to keep in mind that B_2 is an unknown. Its value should be determined such that the currently computed u profile will yield a u_m value that is equal to the given u_m at the entrance. Otherwise the mass is not conserved. Therefore, the flow chart of the code can be constructed as shown in Fig. 17-2.

3-1 Laminar Flows in Two-Parallel-Plate Channels

```
clc; clear % laminar flows in a plate channel
gaB=5000; % an over-relaxation factor or under-relaxation factor
b1=0.02;
new=8.55e-7; rho=1000;
umn=0.02;
Re=umn*b1/new % = 467.84 The flow should be laminar.
B2=3; % guessed first
ny=100; nyp=ny+1; dy=b1/ny; dys=dy*dy;
for j=1:nyp
 y(j)=(j-1)*dy;
end
a(nyp,nyp)=0; b(nyp)=0;
a(1,1)=1;
% 3-point approximation for du/dy=0 at the centerline
a(nyp,ny-1)=.5; a(nyp,ny)=-2; a(nyp,nyp)=1.5;
for iB=1:10
for j=2:ny
 a(j,j-1)=-1; a(j,j)=2; a(j,j+1)=-1; b(j)=dys*B2;
end
```

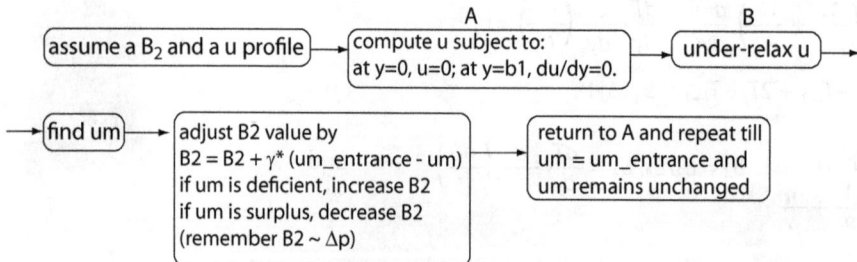

Fig. 17-2   The flow chart of computing u(y) for turbulent flows in planar channels

```
u=a\b';
um=trapz(y,u)/b1 % to find um
B2=B2+ gaB*(umn - um); %
end
plot(y,u)
B2 % = 150.0013
% analytical
B2_analytical = 3*umn/(b1*b1) % = 150
u_center = u(nyp) % = 0.03 check!
```

3-2 Turbulent Flows

See the code in the appendix. The logic is similar to that for laminar flows. The only additional step is to iterate for the u solution since Eq. (8) is highly nonlinear.

3-3 Laminar Flows in Circular Tubes

Considerable literature results have been associated with flows in circular tubes. We feel obligated to at least present the case of laminar flows in circular tubes in the textbook.

The governing equation for u in cylindrical coordinates in fully developed flows can be written as

$$\frac{\mu}{r} \frac{d}{dr}\left(r\frac{du}{dr}\right) = \frac{dp}{dx}, \text{ or} \tag{11}$$

$$-\frac{d^2u}{dr^2} - \frac{1}{r}\frac{du}{dx} = C_1, \text{ where } C_1 = -\frac{1}{\mu}\frac{dp}{dx} > 0.$$

$$-u_{j-1} + 2u_j - u_{j+1} + rr2(j) * (u_{j-1} - u_{j+1}) = C_1 (\Delta r)^2, \tag{12}$$

where $rr2(j) = \Delta r/(2r(j))$, introduced for convenience.

The logic for the remainder of the code is the same as that for the flat-plate-channel case. We just have to be careful when we compute u_m. Do not forget the factor of r in the integral.

See the code in the appendix.

4. Dimples on Golf Balls

So, why are there dimples on golf balls? The answer lies in the aerodynamics around a sphere or an object.

There are two types of drag forces experienced by a flying ball. The first type results in the friction between the airflow and the ball's surface. This surface friction, however, constitutes only a small portion of the entire drag acting on the ball. The major portion of the drag stems from the difference between the low pressure in the wake after the flow separation behind the ball and the high pressure in the stagnation region in front of the ball.

In laminar flows over a sphere, the separation location is nearer to the front than the turbulent counterpart is, as shown in Fig. 17-3. Consequently, the size of the wake for turbulent flows is much smaller than in the laminar flows. Dimples or roughness on the ball's surface cause the turbulent flows to occur.

5. Summary

The governing equation of u(y) for turbulent flows in parallel-plate channels has been derived and solved.

To facilitate our understanding of the treatment of dp/dx term, we also solve the laminar case first.

It is a good opportunity in this lesson to introduce the formulation and the solution of laminar flows in circular tubes.

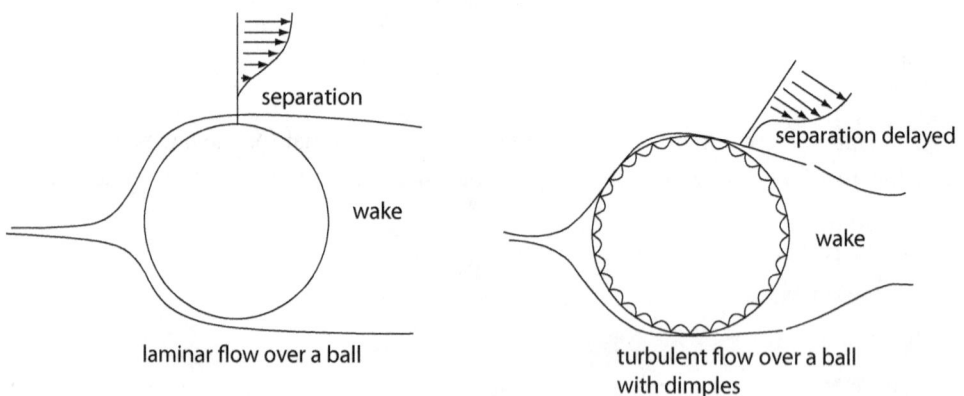

Fig. 17-3   Flows over golf balls with and without dimples

6. Reference

1. E. R. van Driest, On turbulent flow near a wall, *J. Aeronaut. Sci.*, vol. 23, pp. 1007–1011, 1956.

7. Exercise Problems

17-1 Find Nu at Pr = 10 and Re = 4678.4, where Nu = hb_1/k and Re = $u_m b_1/v$.

17-2 Plot $\overline{u'v'}$ versus y for Re = 4678.4.

8. Appendix

A-1 Laminar Flows in Circular Tubes

```
clc; clear
gaC=5000; % an over-relaxation factor or under-relaxation factor
R=0.03;
new=8.55e-7; rho=1000; mu=new*rho;
umn=0.03;
Re=umn*R/new % =1052.6 The flow should be laminar.
C1=100; % guessed first
nr=100; nrp=nr+1; dr=R/nr; drs=dr*dr;
for j=1:nrp
 r(j)=(j-1)*dr; rr2(j)=dr/(2*r(j)+ 1e-10);
end
a(nrp,nrp)=0; b(nrp)=0;
a(1,1)=1.5; a(1,2)=-2; a(1,3)=0.5;
% 3-point approximation for du/dy=0 at the centerline
a(nrp,nrp)=1;
for iB=1:10
for j=2:nr
 a(j,j-1)=rr2(j)-1; a(j,j)=2; a(j,j+1)=-rr2(j)-1; b(j)=drs*C1;
end
u=a\b';
um=2*trapz(r,r.*u')/(R*R) % to find um
C1=C1+ gaC*(umn - um) %
end
```

```
plot(r,u)
C1 % = 266.6505
% analytical
u_center = u(1) % = 0.06 check!
dx=1;
pressure_force = C1*mu*dx*pi*R*R % = 6.4462e-4 N
tauw=mu*(-.5*u(nr-1)+2*u(nr)-1.5*0)/dr; % 3-point approximation
shear = tauw*dx*2*pi*R % = 6.4448e-4 N
```

>>>>>>>>>>

A-2 Fully Developed Turbulent Flows in Two-Parallel-Plate Channels

```
clc; clear
ga=0.4;
gaB=10000; % an over-relaxation factor or under-relaxation factor
b1=0.02;
new=8.55e-7; rho=1000;
umn=0.2;
Re=umn*b1/new % = 4678.4
B2=9200; % guessed first
ny=200; nyp=ny+1; dy=b1/ny; dys=dy*dy; dy2=dy*2;
for j=1:nyp
 y(j)=(j-1)*dy; yplus(j)=y(j)*0.0119/new;
 fac =1-exp(-yplus(j)/33);
 Lv(j)=0.4*y(j)*fac; % van Driest mixing length
 nus(j)=Lv(j)*Lv(j)/new; %initial guess
 u(j)=umn; % initial guess
end
a(nyp,nyp)=0; b(nyp)=0;
a(1,1)=1;
% 3-point approximation for du/dy=0 at the centerline
a(nyp,ny-1)=.5; a(nyp,ny)=-2; a(nyp,nyp)=1.5;
%
for iB=1:35
 for iter=1:30
 for j=2:ny
 dudy(j)=(u(j+1)-u(j-1))/dy2;
 nus(j)=dudy(j)*Lv(j)*Lv(j)/new;
 end
```

```
nus(1)=0; nus(nyp)=nus(ny);
%
for j=2:ny
jm=j-1; jp=j+1;
dnusdy=(nus(jp)-nus(jm))/dy2;
E1=dudy(j)*dnusdy; E2=1+nus(j); B3=(B2+E1)/E2;
a(j,jm)=-1; a(j,j)=2; a(j,jp)=-1; b(j)=dys*B3;
end
q=a\b'; u=ga*q'+(1-ga)*u; % under-relaxation
end
um=trapz(y,u)/b1 % = 0.2 converged to umn
B2=B2 + gaB*(umn - um);
%plot(y,u); hold on
end
%hold off
% post processing
tauw=new*rho*(u(2)-0*.5*u(3))/dy;
utau=sqrt(tauw/rho) % = 0.0119
Cf =2*tauw/(rho*umn*umn); % = 0.00707 ............... very good agreement
Cf_nagaosa = 0.05905*Re^(-.25); % = 0.00714 ........very good agreement
fprintf('%9.5f %9.5f \n', Cf, Cf_nagaosa)
Z=1; dx=1;
shear=tauw*Z*dx % = 0.1414 N ...........................forces balanced
pressure_force=B2*new*rho*dx*b1*Z % = 0.1408 N...forces balanced
```

Lesson 18

Heat Exchangers and Other Heat Transfer Applications

Heat exchangers are important heat exchange devices in industries and daily life. This lesson describes only the fundamentals in simple heat exchangers of the parallel flow type.

Nomenclature

b1 = half height of the channel, m
dy = grid interval, m
e1 = the thickness of the dividing plate, m
kp = thermal conductivity of the plate, W/m-K
LMTD = logarithmic mean temperature difference
NTU = number of thermal units
pec = Peclect number in the cold flow
peh = Peclect number in the hot flow
q_HX = heat transfer from the hot flow to the cold flow
T_h, T_c = T of hot flows or T of cold flows, C
Tw = T at the west point to Tj, C
uc = flow velocity in cold flows, kg/s
uh = flow velocity in hot flows, kg/s

1. Types of Heat Exchangers

In terms of flow directions, there are parallel-flow type and counter-flow type. The numerical analysis of the former is easier, and hence we will focus on it. In terms of the construction, there are finned cross-flows and un-finned cross-flows.

Another type is shell-and-tube heat exchangers shown in Fig. 18-1.

U-tube heat exchanger

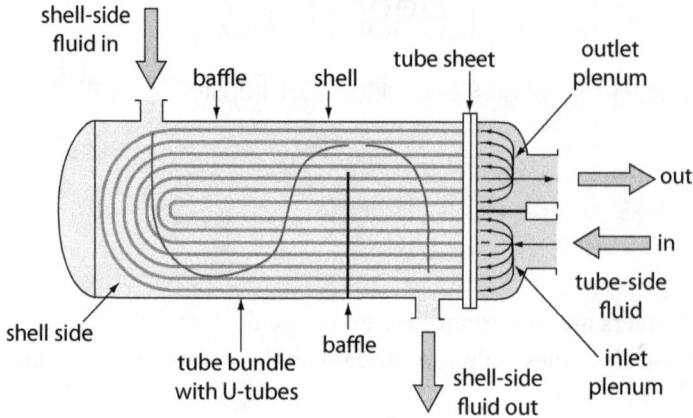

Fig 18-1   A typical shell-and-tube heat exchanger

One of the simplest heat exchangers is a system consisting of two flows divided by a plate. Flows are at two different temperatures and heat transfer takes places through the dividing plate, as schematically shown in Fig. 18-2.

2. A Fundamental Analysis

The governing equation for a typical interior nodal temperature, T3, can be written as:

$$u_c \, (-T_W + T_3)/\Delta x = \alpha \, (T_2 - 2T_3 + T_4)/(\Delta y)^2 \,. \tag{1}$$

The left-hand side of Eq. (1) indicates the net enthalpy carried out by the fluid from the control volume surrounding node 3, with the aid of the upwind finite difference. The right-hand side of Eq. (1) represents the net heat conduction into the control volume.

Other nodal temperatures at 2, 4, 7, 8, and 9 can be derived similarly. Under the assumption that the dividing plate is thin, and $\partial T/\partial y \gg \partial T/\partial x$, the governing equation for T5 can be written as:

$$k(T_4 - T_5)/\Delta y = k_p \, (T_5 - T_6)/e_1 \,,$$

which is the governing equation for T_5. The subscript "p" stands for "plate." The governing equation for T_1 is simply

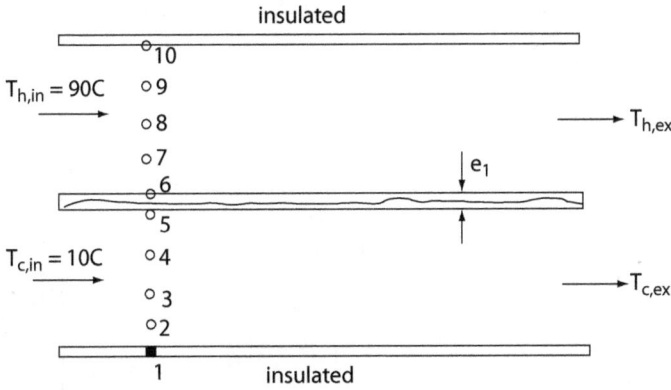

Fig. 18-2   A simple heat exchanger consisting of two planar channels. Heat transfer occurs through the dividing plate.

$$T_1 - T_2 = 0$$

due to insulation.

After finding $T_1, T_2,..., T_{10}$, we update $T_{w1}, T_{w2},..., T_{w10}$ with $T_1, T_2,..., T_{10}$, and continue to march along the flow direction till the exit of the heat exchanger by repeating the same computational procedure. After finding all the nodal temperatures in the computational domain, we can sum up $k_p(T_5 - T_6)/e_1$ at all locations to find the heat exchange between two flows. The numerical result of two temperature distributions of the hot flow and the cold flow is plotted in Fig. 18-3.

3. A Traditional Method to Find Heat Exchange

In the literature [1–3], instead of finding the temperature details as found above, scholars used the Logarithmic Mean Temperature Difference (LMTD), defined as

$$\text{LMTD} = (A - B) / \ln(A/B), \tag{2}$$

where A = T_hot_inlet – T_cold_inlet and B = T_hot_exit – T_cold_exit for parallel flows.

Incidentally, for counter-flows, we should use, instead,

A = T_hot_inlet – T_cold_exit and B = T_hot_exit – T_cold_inlet.

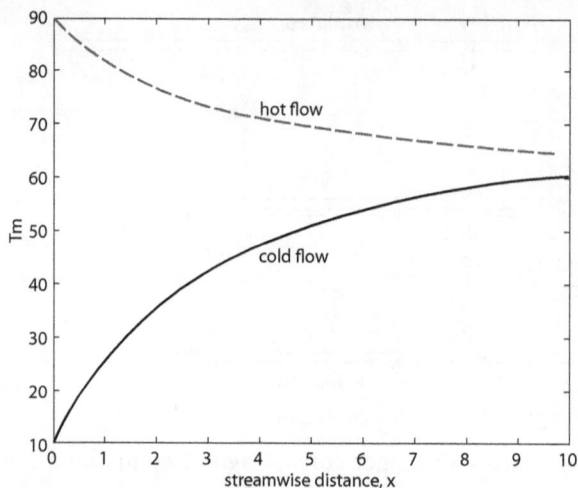

Fig. 18-3   Temperature distributions of the hot flow and the cold flow inside two parallel channels

This value of LMTD can be found first if these four T values are given. It is then used to find the heat exchange rate between two flows as

$$q_s = UA*(LMTD), \tag{3}$$

where $\dfrac{1}{U} = \dfrac{1}{h_c} + \dfrac{e1}{k_P} + \dfrac{1}{h_h}$.

If values of h_c and h_h can be estimated, we can compute and compare the result found in the Matlab code given below with the result found by using Eq. (3).

4. A Matlab Code

```
clc; clear
Z=1; L=10; b1=0.01; dy=b1/2; % channel geometries.
nx=40; nxp=nx+1; dx=L/nx; dys=dy*dy;
e1=0.001; kp=30;
cp=4180; k=0.61; rho=1000; aLf=k/(rho*cp); % fluid properties
Tm_inh = 90; Tm_inc=10; uc=.003; uh=.006; % flow conditions
c1=(kp*dy)/(k*e1); pec=uc*dys/(dx*aLf); peh=uh*dys/(dx*aLf);
%
for j=2:4; Tw(j)=Tm_inc; end
for j=7:9; Tw(j)=Tm_inh; end
```

```
for j=5:6; Tw(j)=0.5*(Tm_inc+Tm_inh); end
Tw(1)=Tw(2); Tw(10)=Tw(9);
% for plotting only
Tmc(1)= Tm_inc; Tmh(1)=Tm_inh; x(1)=0;
%
a(10,10)=0; b(10)=0;
for i=2:nxp
x(i)=(i-1)*dx;
a(1,1)=-1; a(1,2)=1;
for j=2:4
 a(j,j-1)=1; a(j,j)=-(2+pec); a(j,j+1)=1;
 b(j)=-pec*Tw(j);
end
a(5,4)=1; a(5,5)=-(1+c1); a(5,6)=c1;
a(6,5)=c1; a(6,6)=-(1+c1); a(6,7)=1;
for j=7:9
 a(j,j-1)=1; a(j,j)=-(2+peh); a(j,j+1)=1; b(j)=-peh*Tw(j);
end
a(10,9)=1; a(10,10)=-1;
q=a\b'; Tw=q; q' % update the temperatures and march along x
%
Tmc(i)=(q(2)+q(3)+q(4))/3; Tmh(i)=(q(7)+q(8)+q(9))/3;
q_HX(i)=dx*Z*kp*(q(5)-q(6))/e1;
end
plot(x,Tmc, x, Tmh,'r--'); xlabel('streamwise distance, x');
ylabel('Tm'); text(4, 46, 'cold flow'); text(4, 73, 'hot flow');
hold off
%
dT1=Tm_inh - Tm_inc; dT2=Tmh(nxp)-Tmc(nxp);
LMTD= (dT1-dT2)/log(dT1/dT2);
Nu=1.98; %(average of 2.0586 for constant qs and 1.8968 for constant Ts)
hc=Nu*k/b1; hh=hc
U= 1/(1/hc + e1/kp + 1/hh);
qs_literature =-U*Z*L*LMTD % = -1.5883e4 W
%
sum=0;
for i=2:nxp
 sum=sum+q_HX(i);
end
qs_HX = sum % = -9.4587e3 W
```

```
% check the global energy conservation
Gain_cold = rho*3*dy*Z*uc*cp*(Tmc(nxp)-Tmc(1)) % = -9.4587e3 W
Loss_hot = rho*3*dy*Z*uh*cp*(Tmh(nxp)-Tmh(1)) % = -9.4587e3 W
```

5. Comments on the Code

A few comments on the Matlab code can be made below:

(a) We can run the code to find out Nu_cold and Nu_hot.
   See Problem 18-1.

(b) If the problem is of counter-flow type, are we able to modify the code slightly and solve the problem? The answer is "probably not." Let us assume that the hot flow is west-bound. We will have to change the governing equation for the hot-flow temperature to include the east node, not the west node. And the solution of the entire domain must be iterated.

(c) Three quantities should be the same: (1) the energy gained by the cold flow; (2) the energy transferred from the hot flow to the cold flow; and (3) the energy lost by the hot flow. And, indeed, they are all equal 9.4587 kW.

(d) If we set u_h to be equal to u_c, the two Tm (x) curves should be symmetrical to the T=50C line.

(e) As u_h or u_c is increased, the heat exchange rates increase, partly because the mean temperature differences between the cold flow and the hot flow also increase.
   See Problem 18-2.

(f) The code can be slightly modified to compute $\partial T/\partial x$ to assess the validity of neglecting this term in deriving the governing equation (1).
   See Problem 18-3.

6. Other Applications in Heat Transfer

In this lesson, let us take the opportunity to briefly mention other widely studied subjects that are beyond the scope of this textbook. Readers who are interested in these subjects are advised to look for references cited in [3, 4].

6-1 Combustion and Low-Temperature Chemical Reactions

In general, these phenomena will bring source terms or sink terms into the energy equation. Radiation will become important due to high-temperature media and confined walls. Furthermore, species governing equations will be coupled with other governing equations. Fires are typical examples.

You will not be fired or under fire if you study fire and have fire in your heart for her.

6-2 Jets, Plumes, and Wakes

Usually, jets are referred to flows driven primarily by external devices, whereas plumes are those driven by buoyancy forces. Wakes are flows behind an object.

6-3 Optimization

In the subject of optimization, we usually have at least two basic elements: (1) constraint and (2) goal. For example, we are given a 1m-long rope. What is the maximum area we can use this rope to enclose? The length of 1m is the constraint, and the maximum area is the goal.

This subject is particularly important in the design of efficient heat exchangers.

6-4 Porous Media

In porous media, the governing equations have changed. For example, momentum equations often must obey Darcy's law or its variant.

6-5 Radiation with Participating Gases

Since radiation in the form of electromagnetic waves can propagate very long distances, the energy equation T_{ij} for in a typical computational cell will be affected by a chunk of gases located far away, if gases are radiatively participating, as shown Fig. 18-4. In addition, radiation is usually a function of wavelength and solid angle, further complicating the problems.

6-6 Two-Phase Flows

These problems are involved with boiling, condensation, bubble movements, droplet movements, evaporation, and sublimation, among other phase-changing phenomena.

The state of the art in this subject is mostly attempted by experimental and empirical work.

7. Summary

1. In this lesson, we introduced a simple heat exchanger system, and conducted a numerical analysis to find temperatures and fluxes. We can run numerical experiments with the Matlab code provided here.
2. It is generally a good habit to check the validity of the code by examining if the global energy balance is satisfied.
3. Six advanced topics in heat transfer are briefly mentioned.

8. References

1. W. M. Kays and A. L. London, *Compact Heat Exchangers*, 3rd ed., McGraw-Hill, 1984.
2. S. Kakac, A. E. Bergles, and F. Mayinger, *Heat Exchangers*, Hemisphere Publishing, 1981.
3. Tien-Mo Shih, Chandrasekhar Thamire, Chao-Ho Sung, and An-Lu Ren, "Literature Survey on Numerical Heat Transfer (2000-2009): Part I," *Numerical Heat Transfer*, Part A, Vol. 57, pp. 159–296, 2010.
4. Tien-Mo Shih, Martinus Arie, and Derrick Ko, "Literature Survey on Numerical Heat Transfer (2000-2009): Part II," *Numerical Heat Transfer*, Part A, Vol. 60, pp. 883–1096, 2011.

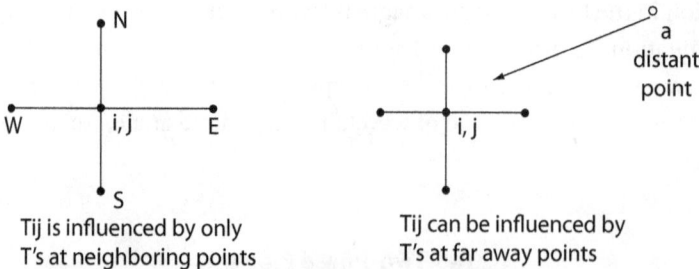

Tij is influenced by only T's at neighboring points

Tij can be influenced by T's at far away points

Fig. 18-4 **Difference between influences on T_{ij} for cases of radiatively participating gases and non-participating gases**

18-1 Use the Matlab code and compute Nu_cold and Nu_hot. Are they functions of x?

18-2 Plot heat exchange rate versus u_c $(= u_h)$ in the range $[0, 0.02 \text{ m/s}]$.

18-3 In deriving Eq. (1), we neglected the heat conduction in x direction, . Find these values for j = 4 at both x = Δx and dx = 0.5L and compare with values of $\partial T/\partial y$, to see if it was acceptable for us to have done such neglecting.

Lesson 19
Radiation (I)

Radiation is the third heat-transfer mode that we will study in the following three lessons. It is an important and broad subject [1]. The scope of our study is limited to wavelength-dependent radiation from (or into) surfaces. Radiation becomes more predominant as the temperature increases.

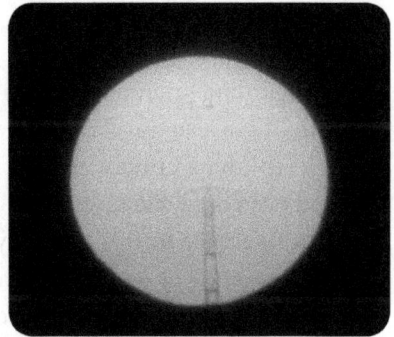

Nomenclature

$E_{\lambda,b}$ = spectral emission from a blackbody surface, $W/m^2\text{-}\mu$

E_b = emission from a blackbody surface, W/m^2

F = fraction of blackbody band emission, defined in Eq. (4)

ε = total emissivity

λ = wavelength, in microns

μ = micron, $10^{-6}m$

σ = Stefan-Boltzmann constant, $W/m^2 - K^4$

1. Fundamental Concepts

Radiation is a phenomenon that involves propagation of electromagnetic waves. In our daily life, we have encountered various terms that span the radiation spectrum, including gamma rays, X-rays, ultraviolet light, visible light, infrared, microwave, TV wave, and radio wave. They are mentioned according to the wavelength in ascending order. The visible light to human eyes, for example, ranges in [0.4 μ, 0.7 μ], as shown in Fig. 19-1.

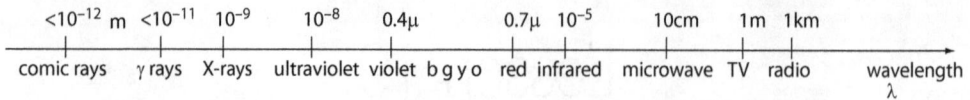

<10^{-12} m	<10^{-11}	10^{-9}		10^{-8}	0.4μ		0.7μ	10^{-5}	10cm	1m 1km	
comic rays	γ rays	X-rays	ultraviolet	violet	b g y o	red	infrared	microwave	TV	radio	wavelength λ

Fig. 19-1   Electromagnetic spectrum in terms of the wavelength

1-1 Main Difference Between Radiation and Convection

Different features between the radiation mode and the convection mode can be identified and listed in the following table:

	radiation	**convection**
(a) medium	not needed	needed
(b) distance	can reach very far	only in the vicinity
(c) speed	speed of light	relatively low
(d) math	integral equations	differential equations
(e) q_s''	$\varepsilon\sigma T_s^4 h$	$(T_S - T_\infty)$

Regarding (a), radiation can travel in vacuum, as the sunlight travels from the sun to earth. It can also travel in air, as we can see the sun and enjoy the sunshine in the winter.

In (b), for example, we can see a candle afar, but can hardly feel it if we place our hand even only one foot away.

In reference to (c), generally, conduction is the slowest mode for the energy front to propagate. For example, it takes about one day for a soil particle at x = 0.5m deep underground to realize that a cold front at -15C has arrived above the ground.

In (d), partly related to (b), the radiation term included in the energy equation governing a heat-transfer phenomenon in a radiatively participating gas may need to be expressed in a multiple integral with respect to distance, wavelength, and solid angle.

1-2 Corresponding Adjectives Used for Radiative Properties Related to Wavelength and Solid Angle

Radiation is usually a function of both wavelength λ and solid angle ω. The nature of wavelength dependence is, in the author's opinion, more interesting and more important. Consideration of solid-angle dependence will inevitably result in integrals containing $\sin(\theta)$, $\cos(\varphi)$, etc., and we may easily become overwhelmed by trigonometry math. Hence, in this textbook we will focus on the former. Regarding the latter, let us simply be aware of the following adjectives:

solid angle	wavelength	
angular	spectral	(for example, angular emissivity will not be studied)
hemispherical	total	
diffuse	gray	

A diffuse surface means a surface that emits the same radiative intensities at different solid angles. A gray surface means a surface that emits the same radiation at different wavelengths.

Angular emissivity is a function of the solid angle; hemispherical emissivity is the average quantity of angular emissivity, and is a constant at a given temperature. These two properties are qualitatively shown in Fig. 19-2. Similar adjectives apply to wavelength dependence.

2. Blackbody Radiation

When we study a subject, often it is beneficial for us to study its ideal cases first. Ideal cases provide us with fundamentals, and are usually simpler for us to handle and understand. Blackbody surfaces are good examples of idealized surfaces.

2-1. Definition of a Blackbody Surface

(a) It will absorb all incidental radiation, regardless of wavelengths. Nothing is reflected or transmitted.

(b) It emits the maximum amount of radiation among all the surfaces at the same temperature. In other words, $\varepsilon = 1$ for blackbody surfaces.

It is worth noting that a black surface is not necessarily a blackbody surface; it merely reflects the black light into our eyes, such as a blackboard or a black dress. Black color to our eyes can be either a specific combination of three primary colors (red, green, and blue) or a result of lacking radiation in the visible light range.

The aperture of a deep hole can be approximated as a blackbody surface.

Guess what: even though the sun looks so bright, it can be approximated as a blackbody surface.

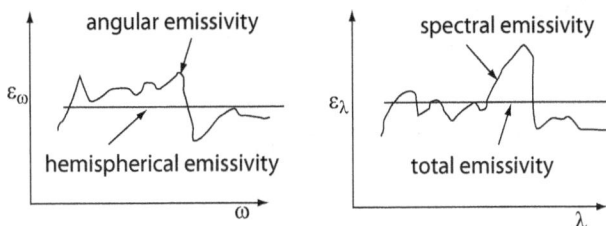

Fig. 19-2   **Typical angular emissivity and spectral emissivity**

2-2. Planck Spectral Distribution

Blackbody emission obeys the following equation

$$E_{\lambda,b} = \frac{C_1}{\lambda^5 \, [\exp(C_2/\lambda T) - 1]} \, , \tag{1}$$

where $C_1 = 3.742e8$ and $C_2 = 1.439e4$.

This equation, known as Planck blackbody emissive spectral distribution and plotted in Fig. 19-3, describes the spectral radiative flux in terms of λ and T. It is derived from quantum mechanics, which is beyond the scope of this course. The unit of this spectral radiative flux is W/m^2-μ.

A few remarks can be made regarding Eq. (1):

(a) The peaks of all the temperature profiles can be connected to yield

$$\lambda_{max} T = 2898, \tag{1a}$$

which is known as Wien's displacement law. Both Max Planck and Wilhelm Wien are Germans, Nobel laureates, and pioneers in radiation.

(b) Solar radiation coincides with the curve at T = 5800K. It can be seen from Fig. 19-4 that, through evolution, our eyes have gradually adapted to capture solar radiation in the vicinity of the peak, which lies in the visible-light range.

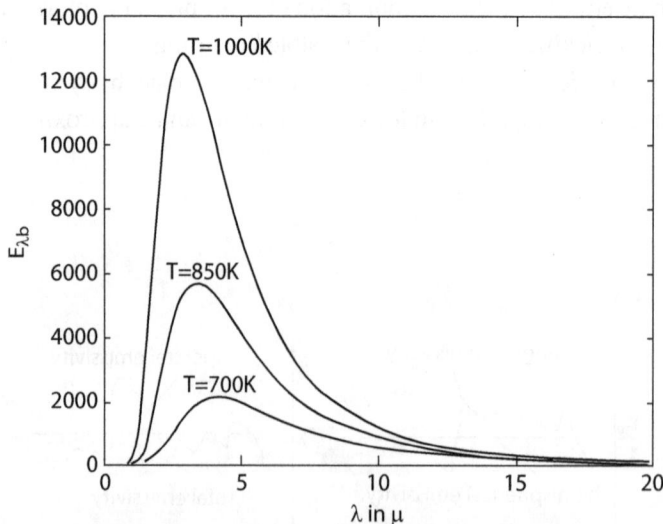

Fig. 19-3 **Blackbody Spectral Emissive Fluxes as a function of wavelength, parameterized in T.**

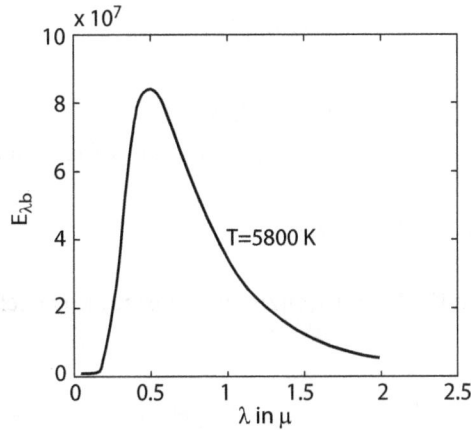

Fig. 19-4   Blackbody Spectral Radiation of the sun, whose apparent T is approximately 5800K

(c) When an iron bar is heated in the furnace to about 800K, it will start to glow into our eyes. The left tail end of the curve at T=700K is at such a low value of $E_{\lambda,b}$ that our eyes cannot detect the radiation.

(d) Let us not mix up reflection with emission. The reason that our eyes can see tree leaves as green is not that they emit green light. Rather, leaves merely reflect the green light into our eyes. The issue of why plants' leaves are green, look green, or reflect green light may be a biologically complicated phenomenon, related to the evolution of leaves' pigment called chlorophyll. For leaves to be able to emit green light, their temperature must be as high as roughly 800K.

(e) The reason we see cats' glowing eyes at night is also due to the reflection of very dim night light by a special reflective layer behind their eyes, called tapetum lucidum. The light we see from their eyes is not emitted by them.

2-3. Stefan-Boltzmann Law

Equation (1) can be integrated over the entire range of wavelength to yield:

$$E_b = \int_0^\infty \frac{C_1}{\lambda^5 \,[\exp\,(C_2/\lambda T)-1]}\, d\lambda = \sigma T^4, \tag{2}$$

where $\sigma = 5.67e\text{-}8$ W/m^2 - k^4 is known as Stefan-Boltzmann constant. It is one of the easiest numbers for us to remember; it consists of four consecutive numbers: 5, 6, 7, and 8.

Example 19-1

Given: a blackbody surface at Ts = 900K, A = $2m^2$.
Find: the radiative heat transfer rate (W) emitted by the surface
Sol:
$q_s = \sigma A T^4 = 5.67 * 2 * 9^4 = 7.44e4$ W.

It is noted that e-8 in the Stefan-Boltzmann constant is canceled with e+8 in $(900)^4$. See Problem 19-1.

3. A Coffee Drinking Tip

On a Sunday morning, we are about to pour some cold cream into hot coffee in the cup, and have a relaxing reading of CNN news. Suddenly, the phone rings. Our classmate is calling us to discuss a heat-transfer home-work problem. We expect the conversation to last 30 minutes. At this instant, we have a very important decision to make; should we:

(1) pour the cream first, then answer the phone, or

(2) answer the phone first, and then return to pour the cream?

The decision is subject to the assumption that we desire to maintain coffee to be as hot as possible.

The governing equation of the coffee temperature can be derived using the lumped-capacitance model as:

$$mc\frac{dT}{dt} = A h (T_\infty - T) + A\alpha \sigma T_\infty^4 - A\epsilon \sigma T^4. \tag{3}$$

In establishing Eq. (3), we have assumed that heat transfer takes place only via the coffee surface. The bottom and the circumferential area of the coffee cup are well insulated. Also, for simplicity, we will further assume $\varepsilon = \alpha$. The following Matlab code is for us to run some numerical experiments. Results are shown in Fig. 19-5. Based on our computation, we definitely would prefer to hurry up and pour the cream first before answering the phone.

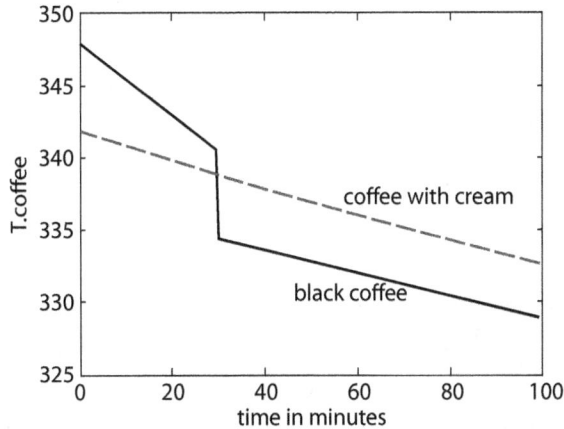

Fig. 19-5   Temperatures of black coffee and coffee with cream vs. time

```
%%
clc; clear
dt=20; h=4; A=7e-3; m=0.2; c=4180;
sig=5.67e-8;
Tinf=298; ntm=300; tm(1)=0;
Tp=348; eps=.9; T(1)=Tp; % black coffee
for itm=2:ntm
tm(itm)=(itm-1)*dt;
if(tm(itm) ==1800); Tp=Tp-6; eps=0.1; end;
Conv=(A*h/(m*c))*(Tinf-Tp);
Rad=(A*eps*sig/(m*c))*(Tp^4-Tinf^4);
% using the simple explicit method
S1=Conv-Rad; T(itm)=dt*S1+Tp; Tp=T(itm);
end
plot(tm/60, T); hold on
xlabel('time in minutes'); ylabel('T.coffee');
%
Tp=342; eps=.1; T(1)=Tp; % coffee with cream, Tp is 6C lower than the black coffee
case
for itm=2:ntm
Conv=(A*h/(m*c))*(Tinf-Tp);
Rad=(A*eps*sig/(m*c))*(Tp^4-Tinf^4);
S1=Conv-Rad; T(itm)=dt*S1+Tp; Tp=T(itm);
end
plot(tm/60, T, 'r--')
text(51, 331, 'black coffee'); text(58, 337, 'coffee with cream')
```

hold off
% Let us sketch T_black vs. time and T_cream vs. time qualitatively
See Problem 12-2.

4. Fractions of Blackbody Emission

Instead of over the entire range of wavelength, frequently we are interested in finding a portion of the emissive radiation between λ_1 and λ_2, shown in Fig. 19-6. Such a fraction, F, can be derived as:

$$F(\lambda_1 T - \lambda_2 T) = \int_{\lambda_1}^{\lambda_2} E_{\lambda,b}\, d\lambda / \sigma T^4$$

$$= \int_0^{\lambda_2} E_{\lambda,b}\, d\lambda / \sigma T^4 - \int_0^{\lambda_1} E_{\lambda,b}\, d\lambda / \sigma T^4$$

$$= F(0 - \lambda_2 T) - F(0 - \lambda_1 T). \tag{4}$$

Values of these fractions can be computed by using the trapezoidal rule and running a Matlab code given in the appendix. The table of fractions is also presented in the appendix. It can be constantly referred to when we need to find total radiative properties in later sections.

Example 19-2
Given: a blackbody plate, Ts = 2000K
Find: radiative flux emitted from the plate within the range of $0 < \lambda < 2\mu$

Sol:
$q_b = 5.67\text{e-}8 * (2000)^4 = 907,200 \ W/m^2.$
From the fraction table, we find F = 0.4805 at $\lambda T = 4000$.
So, q (within the range 0 to 2) = $q_b * F = 436.25 \ kW/m^2.$

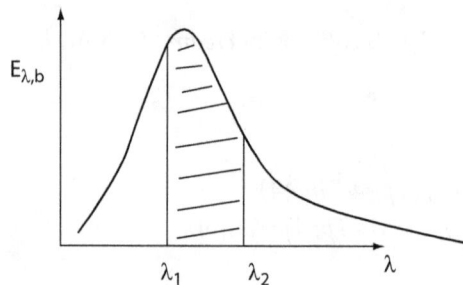

Fig. 19-6   A fraction of blackbody emission between two arbitrary λ's

See Problems 19-3 and 19-4.

In the next lesson, we will study various radiative properties. The knowledge of the band emission will help us to calculate them without having to conduct numerical integrations every time.

5. Summary

In this lesson, we have learned:

(a) fundamental concepts regarding blackbody radiation,
(b) a daily-life problem of coffee drinking with or without cream,
(c) fractions of blackbody emission.

6. References

1. Robert Siegel and John R. Howell, *Thermal Radiation Heat Transfer*, McGraw-Hill, 1972.

7. Exercise Problems

19-1 Which of the following statements regarding Planck's Emission Spectral Distribution is true?
   (a) The area underneath the curve of T=1000K in Planck's Emission Spectral Distribution is a function of the wavelength.
   (b) The peak values of the curves shown in Planck's Emission Spectral Distribution are proportional to the locations, λ_{max}. The higher the peak, the larger the λ_{max}.
   (c) If the area underneath the curve is divided into two portions, one left to the peak and the other right to the peak, then the right area is the same as the left area for all the curves.
   (d) The area underneath the curve is a function of T, but the fraction is a function of the product λT only.

19-2 Find T_coffee_with_cream vs. time using the implicit method and the official Newton-Raphson method.

19-3 What is the percentage of the visible-light range to the entire spectrum of the sunlight?

19-4 Given Ts = 2000K. What is the percentage of E_b left to the peak?

A-1 Finding the Value of Stefan-Boltzmann Constant

```
clc; clear % finding the value of sigma (Stefan-Boltzmann constant)
 % and plot Eb vs. Lambda
c1=3.742e+8; c2=1.439e+4;
L=200; nx=40000; dx=L/nx; nxp=nx+1; % take Lambda between 0 and 2000
for iT=1:3
T=700+(iT-1)*150;
for i=1:nxp
Lam(i)=1e-6+(i-1)*dx; % avoid division by zero.
L5=Lam(i)^5; cLT=c2/(Lam(i)*T); ELb(i)=c1/(L5*exp(cLT)-L5);
end
area=trapz(Lam, ELb);
sig=area/T^4; % = 5.667e-8
plot(Lam(1:4000), ELb(1:4000)); hold on
text(3, max(ELb)+300, ['T=', num2str(T), 'K'])
end
xlabel('\lambda in \mu'); ylabel('E_\lambda_,_b')
hold off


>>>>>>>>>>


clc; clear % finding Fractions of Band Emissions
c1=3.742e+8; c2=1.439e+4; sig=5.67e-8;
L=6; nx=6000; dx=L/nx; nxp=nx+1; % take Lambda between 0 and 6
T=2000;
for i=1:nxp
Lam(i)=1e-6+(i-1)*dx; % avoid division by zero.
L5=Lam(i)^5; cLT=c2/(Lam(i)*T); ELb(i)=c1/(L5*exp(cLT)-L5);
end
for i=501:100:nxp
area=trapz(Lam(1:i), ELb(1:i));
F =area/(sig*T^4); LT=Lam(i)*T;
fprintf('%9.0f %9.4f \n', LT, F)
end
```

A-2 Table 19-1 Fractions of Blackbody Emission

λT	F(0-λT)	λT	F(0-λT)	λT	F(0-λT)
1000	0.0003	4800	0.6071	8600	0.8774
1200	0.0021	5000	0.6333	8800	0.8837
1400	0.0078	5200	0.6575	9000	0.8895
1600	0.0197	5400	0.6799	9200	0.8950
1800	0.0393	5600	0.7006	9400	0.9001
2000	0.0666	5800	0.7197	9600	0.9049
2200	0.1008	6000	0.7373	9800	0.9094
2400	0.1401	6200	0.7537	10000	0.9137
2600	0.1829	6400	0.7688	10200	0.9177
2800	0.2277	6600	0.7827	10400	0.9214
3000	0.2730	6800	0.7956	10600	0.9249
3200	0.3178	7000	0.8076	10800	0.9282
3400	0.3614	7200	0.8187	11000	0.9314
3600	0.4033	7400	0.8290	11200	0.9343
3800	0.4430	7600	0.8386	11400	0.9371
4000	0.4805	7800	0.8475	11600	0.9397
4200	0.5156	8000	0.8558	11800	0.9422
4400	0.5484	8200	0.8635	12000	0.9446
4600	0.5789	8400	0.8707	12200	0.9468

At $\lambda T = 50000$, $F(0-\lambda T) \approx 1$. For most practical purposes, let us simply interpolate F values if they are beyond the range listed above.

Lesson 20

Radiation (II)

In this lesson, four important radiative properties, emissivity ε, absorptivity α, reflectivity ρ, and transmissivity τ will be learned. In addition, problems related to energy balance over a plate will be studied.

Nomenclature

E = radiative flux emitted by the plate

f_1 = the fraction of solar radiation received by earth, 0.0021%

G = incoming radiative flux arriving at the plate

m = the mass of the plate

α = absorptivity (do not mix it up with the thermal diffusivity)

ε = emissivity

ρ = reflectivity (do not mix it up with the density)

τ = transmissivity (do not mix it up with the shear stress)

1. Emissivity

Unlike absorptivity, reflectivity, and transmissivity, the emissivity is the only radiative property among the four that is unrelated to the external incoming radiative fluxes. It is a function of the wavelength and the temperature of the emitting source, such as the one inside the pumpkin.

First, let us define the spectral emissivity as

$$\varepsilon_\lambda = E_\lambda(\lambda, T) / E_{\lambda b}(\lambda, T), \tag{1a}$$

which immediately yields

$$E_\lambda(\lambda, T) = \epsilon_\lambda E_{\lambda,b}(\lambda, T).$$

Usually, the equation above is not too interesting to us. But it can be used for deriving the total emissivity as shown by:

$$\varepsilon = \frac{E(T)}{E_b(T)} = \frac{\int_0^\infty E_\lambda(\lambda, T)\, d\lambda}{\sigma T^4} = \frac{\int_0^\infty \epsilon_\lambda E_{\lambda,0}(\lambda T)\, d\lambda}{\sigma T^4}. \tag{1b}$$

It can be seen that, once we have managed to find the total emissivity, we can readily find the radiative flux emitted by the surface by

$$E(T) = \varepsilon\, \sigma T^4. \tag{2}$$

Therefore, in general, to solve a radiation problem, we will be first given a spectral emissivity distribution. Then, we will find the total emissivity with Eq. (1b). And finally we will use this value to find the emitted radiative flux with Eq. (2).

Example 20-1

Given: Ts = 1000K, and the spectral emissivity shown in Fig. 20-1.
Find: the total emissivity and the emitted flux

Sol:
Judging from the graph, we can divide the integration range into three parts:

(I) = 0 → 4, (II) = 4 → 6, (III) 6 → ∞
Since F_(0 – 4*1000) = 0.4805 and F_(0 – 6*1000) = 0.7373, we obtain
ε = 0.9*0.4805 + 0.3*(0.7373 – 0.4805) + 0 = 0.5095.
$q_s'' = \varepsilon\sigma T_s^4 = 0.5095*5.67*(10)^4 = 2.89\mathrm{e}4 \ W/m^2.$

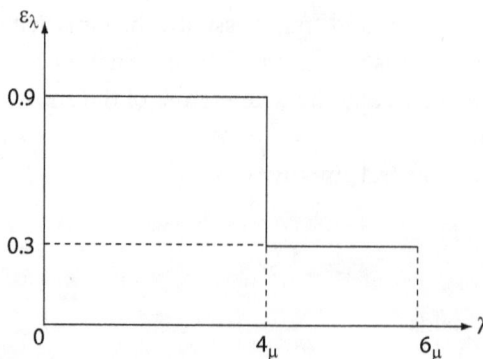

Fig 20-1   Spectral emissivity distribution

Example 20-2

Given: Ts = 1000K.

Find: the graph that is associated with the largest total emissivity among the three, shown in Fig. 20-2.

Sol: Without having to rely on detailed numerical integrations, we already know by intuition that graph (b) should yield the highest total emissivity, simply because the peak of the spectral emissive flux curve should be at $\lambda \approx 3\mu$, based on Wien's displacement law. In graph (b), the integration range covers 3μ well.

Even though the integration range includes 3μ, too, all the emission curves drastically decrease as decreases in the vicinity of $\lambda=0$. So, graph (c) cannot compete with the middle one.

If we do not have sufficient confidence in our prediction, we can actually go through the calculations, of which the results are shown below as:

total emissivity in (a) = 0.8*(0.8076 − 0.4805) = 0.2617.
total emissivity (middle) = 0.8*(0.4805 − 0.0003) = 0.3842…the highest, as expected.
total emissivity in (c) = 0.8*(0.2730 − 0) = 0.2184.

2. Three Other Radiative Properties

As in the case of the spectral emissivity, let us first define the spectral absorptivity as

$$\alpha_\lambda = G_{\lambda a} / G_\lambda \tag{3a}$$

where G stands for the incoming radiative flux, and the subscript "*a*" stands for "absorbed." Consequently, we can obtain the total absorptivity as

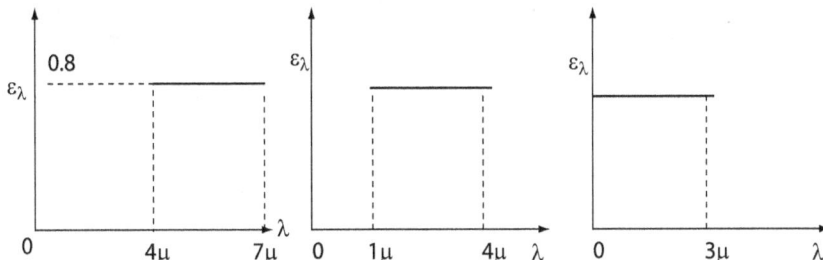

Fig. 20-2   **Choose the maximum total emissivity by intuition**

$$\alpha = \frac{G_a}{G} = \frac{\int_0^\infty G_{\lambda a}\, d\lambda}{\int_0^\infty G_\lambda\, d\lambda} = \frac{\int_0^\infty \alpha_\lambda G_\lambda\, d\lambda}{\int_0^\infty G_\lambda\, d\lambda} \qquad (3b)$$

This equation can be applied to finding the reflectivity and the transmissivity similarly. Let us skip the routines, and simply write the final results:

$$\rho = \int_0^\infty \rho_\lambda G_\lambda\, d\lambda \big/ \int_0^\infty G_\lambda\, d\lambda \qquad (4)$$

and

$$\tau = \int_0^\infty \tau_\lambda G_\lambda\, d\lambda \big/ \int_0^\infty G_\lambda\, d\lambda. \qquad (5)$$

3. Solar Constant and Effective Temperature of the Sun

Prior to some analyses of radiation-related global-warming problems, it is beneficial for us to acquire and estimate basic information about the sun. Figure 20-3 shows the geometry of the sun-earth system.

Q1: How far is the sun away from us?
Answer: Solar Distance = R2 = 93 million miles = twice the distance if we drive 70 miles/hr for 70 years, 24 hours a day and 365 days a year nonstop.

Q2: How hot is the sun?

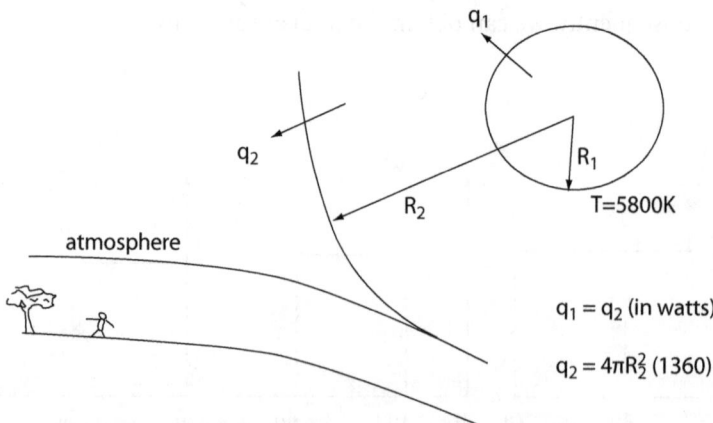

Fig. 20-3   Schematic geometry of the sun-earth system

Answer: very very hot, about 1.6e7K (incidentally, the melting point of diamond is "only" 3820K under regular pressures). But for our purpose of learning thermal radiation, it is sufficient for us to know that:

(a) At R1 = 0.428 million miles away from the center of the sun (or on the sun's "surface"), the sun's radiation is equivalent to the emission from a blackbody plate at Ts ≈ 5800K.

(b) At R2 = 93 million miles away from the sun (or above the atmosphere on earth), the sun's radiation is equivalent to the emission from a blackbody plate at Ts ≈ 393.5K. Or it is equal to 1360 W/m^2, known as the solar constant.

(c) Energy conservation dictates that, as shown in the figure, q1 = q2, where

$$q1 = 4\,\pi R_1^2\, q_1'' = 4\,\pi R_2^2\, q_2'' = q2.$$

Q3: Out of the total sun's emission, what is the percentage, f_1, received by earth?
Answer: f_1 = 1360 / [σ (5800)^4] = 2.1e-5 = 0.0021%

Example 20-3

Let us check if q1 = q2.
Given: data specified in the Matlab code below, and the figure above depicting R1 and R2

```
clc; clear
sig=5.67e-8;
T_eff =5800; % in K
q_sun = sig*T_eff^4 % = 6.4165e7 W/m^2
solarc=1360; % in W/m^2
R2 = 93e6*1600; R1 = 6.851e8; % in m
f1 = solarc/q_sun % = 2.12e-5
%
q1 = 4*pi*R1^2*q_sun % = 3.7845e26 W
q2 = 4*pi*R2^2*solarc % = 3.7840e26 W

% Indeed, they are equal.
```

Example 20-4

Given: the spectral transmissivity distribution of a plate as shown in Fig. 20-4
Find: τ exposed to a blackbody heat source at T_source = 3000K

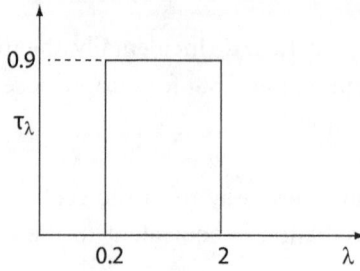

Fig 20-4   Spectral transmissivity distribution

Sol:

First, we need to pay attention to the phrase, "with respect to the heat source at T_ source." To find τ, ρ, or α, we must always mention the source of the incoming radiative flux. If we neglect mentioning the source, the statement itself, "Find τ", is meaningless.

$\tau = 0.9*[\, F(0\text{-}2*3000) - F(0\text{-}0.2*3000)] = 0.9* (0.7373 - 0) = 0.6636.$

4. Gray Surfaces

Gray surfaces are defined as surfaces whose radiative properties are independent of the wavelength. Hence, for gray surfaces, all spectral properties are equal to total properties, and are constants.

If G is the incoming radiation to a plate, then it follows that, in reference to Fig. 20-4a,

G = absorbed amount + reflected amount + transmitted amount.

There cannot be any other quantities that are not accounted for. Hence, if we divide both sides by G, we will obtain

$1 = \alpha + \rho + \tau.$ (6a)

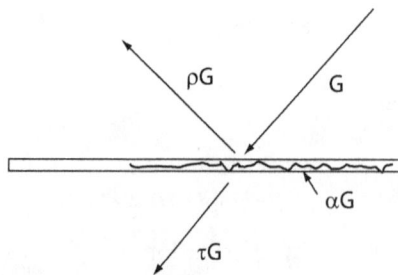

Fig. 20-4a   Incoming radiation, G, is equal to the combination of 3 components.

If the plate is opaque ($\tau = 0$), then

$$1 = \alpha + \rho. \tag{6b}$$

Spectral properties also follow Eqs. (6a, 6b). For a given surface exposed to a known source, if two spectral properties are prescribed, and if both Ts and T_source are specified, we should be able to find the remaining two.

For example, if α_λ distribution and ρ_λ distribution are known, we also know ε_λ immediately according to Kirchhoff's law (to be decribed next). After finding α, ρ, and ε, we can readily find τ from Eq. (6a).

Example 20-5

Consider a plate exposed simultaneously to two heat sources as shown. q1 = 907,200W; q2 = 56,700W. The fractions received by the plate are f1 = 0.62% and f2 = 10%, respectively, such that f1*q1 = f2*q2 = 5,670W, as shown in Fig. 20-5. Furthermore, let us assume that the plate absorbs very little impinging radiation.

Find: two reflectivities due to two different sources, also in reference to Fig. 20-4 for spectral transmissivity.

Sol: $\tau_1 = 0.9*(0.4805 - 0) = 0.4325$;
$\tau_2 = 0.9*(0.0666 - 0) = 0.0599$.

Hence, $\rho_1 = 1 - 0.4325 = 0.5675$.
$\rho_2 = 1 - 0.0599 = 0.9401$.

Note that, even though the arriving fluxes from two different sources are of the same amount, the reflectivities are different.

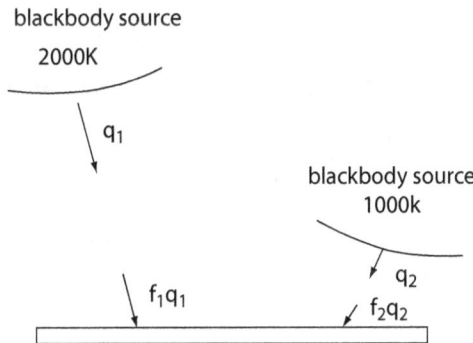

Fig. 20-5   Two reflectivities from two different heat sources

This law states that, for most materials on earth and for most practical purposes, we have

$$\varepsilon_\lambda = \alpha_\lambda. \tag{7}$$

Non-rigorously, we can argue that, if Eq. (7) were not true, we could perceivably manufacture or find a certain type of material whose ε_λ is, for example, 0.1, and α_λ is 0.9 at $\lambda = 0.5\mu$. We then place it within an enclosure at T = 300K. As time elapses, the material will enjoy receiving more energy than it emits. Consequently, the temperature of this material will continue to increase till $0.1 * T_s^4 = 0.9 * 300^4$, or Ts = 519.6K. This phenomenon violates the Second Law of Thermodynamics.

This issue can be logically revisited after we have studied Example 20-13.

A rigorous proof of Kirchhoff's law is beyond the scope of this textbook. It is sufficient for us to keep in mind that a spectral emissivity distribution can be used as the spectral absorptivity distribution, and vice versa.

Equation (7), by no means, automatically implies that

$$\varepsilon = \alpha. \tag{8}$$

There are two following conditions, under one of which Eq. (8) is valid:

(a) if the surface is gray, or
(b) if the source is a blackbody, and is at the same temperature as the plate temperature.

Under condition (a), Kirchhoff's law will automatically imply Eq. (8). Under condition (b), G_λ in Eq. (3b) will become the same as $E_{\lambda,b}$ in Eq. (1b).

Example 20-6

Revisit Example 20-1. The plate is also exposed to a heat source at T_source = 2000K. Find: α

Sol: From the table of blackbody emission fractions, we find
F(0 – 4*2000) = F(0 – 8000) = 0.8558; F(8000 – 12000) = 0.9446 – 0.8558 = 0.0888.
Hence, α = 0.9*0.8558 + 0.3*0.0888 = 0.7969.

In Example 20-1, we found ε = 0.5095. This problem is a good example demonstrating that ε ≠ α. If T_source = 1000K also, then ε = α = 0.5095.

6. Energy Balance over a Typical Plate

Consider an earth-bound plate, emitting radiation, receiving radiation from a radiative heat source, and is exchanging energy with the surrounding air, as shown in Fig. 20-6.

Taking energy conservation over the plate, we have

$$m\,C_v\,\frac{dT}{dt} = h\,A\,(T_\infty - T) + A\,(G - \rho G - \tau G) - A\epsilon\sigma T^4. \tag{9}$$

It is worth noting that the absorbed component is not included in Eq. (9). It should not be included, because the absorption activity takes place inside the control volume, not across the boundary of the control volume.

Equation (9) can be simplified into

$$m\,C_v\,\frac{dT}{dt} = h\,A\,(T_\infty - T) + A\,\alpha\,G - A\epsilon\sigma T^4 \tag{10}$$

which is clearly more preferable than Eq. (9), as it contains fewer terms. Equation (10) will be used in some of the following examples.

See Problems 20-1, 20-2, and 20-3.

7. Greenhouse Effect (or Global Warming)

A slab of semi-transparent material can act as a filter, allowing energy at certain temperatures to pass, but blocking energy at certain temperatures. If such material is some types of glass, the phenomenon is known as Greenhouse Effect. If such material is the layer of CO_2 hanging over earth, we may be speaking of Global Warming.

Example 20-7

Given: a piece of glass, solar constant ($=1360$ W/m^2), with the spectral transmissivity shown in Fig. 20-7.

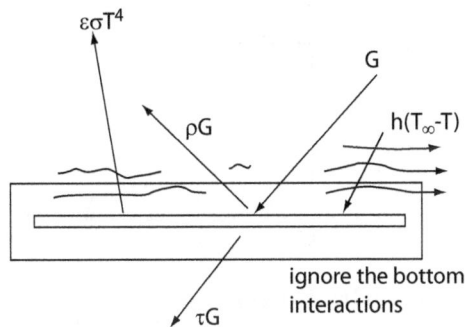

Fig. 20-6   Energy balance over a plate, including all energy components

Find: (1) to the solar radiation, (2) transmitted radiative flux
Sol:
From the table of emission fractions, we obtain

F(0 – 2*5800) = 0.9397; F(0 – 0.2*5800) ≈ 0.0019.

$$\tau = \frac{\int_0^\infty t_\lambda G_{\lambda a}\, d\lambda}{\int_0^\infty G_\lambda\, d\lambda} = \frac{\int_0^\infty \tau_\lambda f_1 E_{\lambda,b}\, d\lambda}{\int_0^\infty f_1 E_{\lambda,b}\, d\lambda} = \frac{\int_0^\infty \tau_\lambda E_{\lambda,b}\, d\lambda}{\int_0^\infty E_{\lambda,b}\, d\lambda}$$

= 0.9*(0.9397 – 0.0019) = 0.844.

$$q''_{tr} = \tau * (1360) = 1147.8 \text{ W}/m^2.$$

If we repeat the calculations for τ with respect to a source at T ≈350K (like the interior of a car under the sunlight in the summer), we will find that τ ≈ 0, suggesting that thermal energy is trapped inside the car.

Example 20-8

Consider the spectral transmissivity of CO_2 layer in the atmosphere shown in Fig. 20-8. Are we able to speculate why earth is warming up based on this figure?
Sol: The total transmissivity with respect to the sun can be calculated as

$$\tau(T_{sun}) = 0.6\, F(0 - 2.7*5800) = 0.6*0.971 = 0.5826.$$

Assuming the average temperature of the earth is approximately 300K, we also can find

$$\tau(T_{earth}) = 0.6\, F(0 - 2.7*300) = 0.6*\text{small} \approx 0.00001.$$

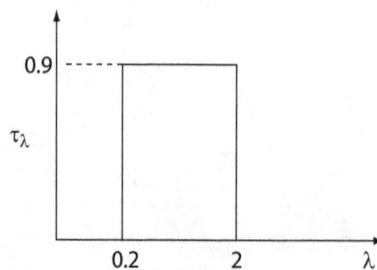

Fig. 20-7   Special transmissivity distribution

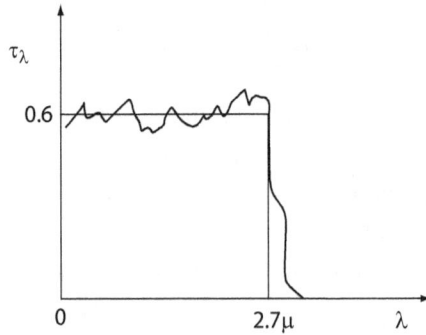

Fig. 20-8   The approximate transmissivity of CO_2 layer

It was estimated that, in the last 100 years, the earth's average surface temperature increased by about 0.8 °C [1]. In analogy, it seems that we earthlings know to take from our sun, but do not reciprocate, pretty much the same as we children know to take from our parents abundantly, but reciprocate to them only very scarcely.

8. Steady-State Heat Flux Supplied Externally by Us

For steady-state problems, the energy balance over a control volume can be simply expressed by

in – out = 0.

Since q"_supplied should be considered as an energy component into the control volume, we can modify the equation above into

q"_supplied = out – in.   (11)

Example 20-9

Given: gray opaque surface placed in the outer space, Ts = 900K, ε = 0.4, h = 0, T_source = 1000K (blackbody)
Find: q"_supplied

Sol:
q"_supplied * A = 2*A* ε * σT_s^4 – A* α* σ*(1000)^4,

or q"_supplied = 2* 0.4 * 5.67*(9)^4 – 0.4* 5.67*(10)^4 = 7080.7 W/m^2.

In Eq. (11), there are two unknowns q"_supplied and Ts. Since we have only one equation, we can afford to have only one unknown. One of the two unknowns must be given. Usually, when Ts is not given, the problem becomes more difficult, because we may need to solve a nonlinear algebraic equation.

In addition, if (1) Ts is not known, (2) the emission from the plate is involved in the energy balance, and (3) the plate is non-gray, then the problem becomes even more cumbersome, because the total emissivity, ε, will be a highly implicit function of Ts. We are unable to find derivatives explicitly, rendering the Newton-Raphson method awkward to use.

Example 20-10

Consider a gray opaque surface placed in the outer space with $\varepsilon = 0.4$ and h = 0. A blackbody heat source at T_source = 1000K is very close to the plate, such that all the emitted energy will arrive at the plate. The plate is in steady state.

Find: Ts

Sol: We expect Ts to be less than 900K.

$$0 = 2*A* \varepsilon *\sigma* T_s^4 - A* \alpha* \sigma*(1000)^4$$

or Ts = $1000* (0.5)^{1/4} = 840.9$ K.

10. Find Steady State Ts Numerically

In Example 20-10, Ts can be found analytically because there are only two terms present in the governing equation. If there exist more than two terms, then an analytical procedure may no longer be possible.

Example 20-11

Given: Let us place the plate on the earth ground, h = 20 W/m^2- K, $T_\infty = 300$K, with the bottom side insulated.

Find: Ts

Sol:
$$0 = \varepsilon\sigma T_s^4 - \alpha \sigma (1000)^4 + h (T_s - T_\infty) \tag{12}$$
or,

$$0 = 0.4^* \, 5.67e\text{-}8T_s^4 - 0.4^*5.67e4 + 20^*(T_s - 300),\qquad\qquad(12a)$$

which is a nonlinear equation containing Ts as the only unknown. Performing trials and errors, we finally obtain Ts = 847.8K.

We can also use the Newton-Raphson method to solve Eq. (12a) officially. Let us linerarize

$f(x) \approx f(xb) + f'(xb)\,^*(x - xb)$ by Taylor's series expansion.

A simple Matlab code solving Eq. (12a) is given below.

```
clc; clear
sig = 5.67e-8;
% 0 = 0.4* sig* Ts^4 - 0.4*5.67*e4 + 20*(Ts - 300)
% Ts^4 = c1 + c2*Ts, c1 = -3*Tb^4; c2 = 4*Tb^3;
% 0 =c3*(c1 + c2*Ts)-c4 + 20*Ts; c3=0.4*sig; c4 =0.4*5.67*e4+6000;
Tb=700; % initial guess
for iter =1:5
c1 = -3*Tb^4; c2 = 4*Tb^3; c3 =0.4*sig; c4 =0.4*5.67e4+ 6000;
T = (c4-c1*c3)/(20+c2*c3);
Tb = T; % update Tb
fprintf('%4.0f %9.4f \n', iter, T)
end
```

iteration	Ts
1	880.6550
2	849.2410
3	847.8952
4	847.8929
5	847.8929

11. Find Unsteady Ts Not Involved with the Spectral Emissivity

The complexity of the problems gradually increases. We now turn our attention to problems that are time dependent.

Example 20-12

Let us revisit the coffee cooling problem presented in Section 19 – 3.
Find: time for the black coffee to cool down by 5K.

Sol:

Looking up the computed chart, we can estimate the time required is about five minutes. The purpose of this example is threefold:

(a) emphasize the importance of coffee-cooling problem, which constitutes a typical daily-life event involving combined convection and radiation.
(b) mention that, in computing the external incoming radiation, we seemed to have assumed the ceiling to be a blackbody, which may not qualify so at all.
(c) prepare us for the next example.

See Problem 20-4.

12. Find Unsteady Ts Involved with the Spectral Emissivity

Why is it such a big issue when ε_λ is involved? The following example will provide the explanations. In addition, this section may also demonstrate the validity of Kirchhoff's law.

Example 20-13 (20-12 modified)

Consider the cooling of non-gray coffee whose spectral emissivity is shown in Fig. 20-9. Everything else is the same as in the preceding example. Ignore the radiation from the surroundings.

Find: time for the black coffee to cool down by 5K.
Sol:

$$mc_v \frac{dT}{dt} = -hA\,(T - T_\infty) - A\varepsilon\sigma T^4.$$

But it is noted that ε is a very implicit function of T. Prior to solving the differential equation, we may desire to find an approximate polynomial expression such as:

Success is a highly nonlinear equation of our effort, our vision, and luck.

$$\varepsilon(T) = a + bT + cT^2 + dT^3.$$

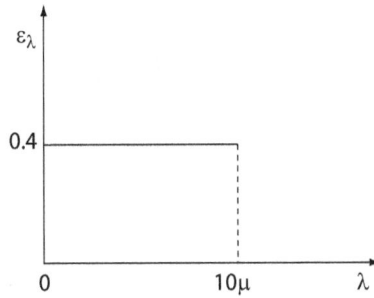

Fig. 20-9   Spectral emissivity distribution of coffee

Then, at T = 400, F[0-(10)(400)] = 0.4805 → ε = 0.192
at T = 350, F[0-(10)(350)] = 0.38 → ε = 0.152
at T = 300, F[0-(10)(300)] = 0.2730 → ε = 0.109
at T = 250, F[0-(10)(250)] = 0.16 → ε = 0.064.

From these four data points, we can determine a, b, c, and d.

If we do not find the emissivity as an approximate function of T *a priori*, it will be quite cumbersome for us to solve T(t) in a continuous fashion.

See Problem 20-5, which is related to Kirchhoff's Law.

13. Find Unsteady Ts with Parameters Being Functions of Wavelength and Time

Rail buckling is induced by thermal stress, and has been the number-one cause of train accidents in the nation. If we are the technical consultants for a rail-buckling project sponsored by DOT, what are our thoughts regarding the thermal aspects of this rail-buckling problem?

Fig. 20-10   The side view of a typical rail under weathered conditions

Thought 1: We may attempt to take energy balance over a section of the rail, of which the side view is shown in Fig. 20-10. The governing equation for T of the rail can be written as:

$$mc_v \frac{dT}{dt} = -h\,A\,(T - T_\infty) - A\epsilon\sigma T^4 + A\,\alpha\,G\,(t), \tag{13}$$

where G(t) includes the daily solar radiation and the emission from the hot ballast underneath the rail.

Thought 2: In Eq. (13), c_v, ϵ, and α are all dependent on the material of the rail structure, which may be affected by thermal stress, which in turn is affected by T(t) of the rail. Furthermore, *h* is not known *a priori*. So, the problem may be a complicated conjugate problem.

Thought 3: In Eq. (13), G(t) is definitely a function of t. However, h, ϵ, and α may likely be functions of t as well. Their values at locations 1, 2, 3, 4, and 5 are clearly different, and will likely vary between sunrise and sunset.

Thought 4: Equation (13) does not consider the space dependence for T. Is such a lumped-capacitance model adequate? If the value of h is small, perhaps the lumped-capacitance model is acceptable. What if the weather becomes windy, rainy, or snowy?

Thought 5: Should we be concerned about the spectral dependence of ϵ and α? This concern will additionally complicate the already-complicated problem.

Thought 6: How is the thermal aspect of the problem coupled with the stress aspect? After all, it is the non-uniform stress distribution that causes the buckling. Is such a coupling one way [T(t, x, y, z) affects stresses, but stresses do not affect T(t, x, y, z)] or both ways?

It is gratifying to know, however, that what we have learned in radiation can be possibly applied to generating some relevant thoughts associated this rail-buckling problem.

14. Summary

This lesson has presented us the following topics:

(a) using the table of blackbody emission fractions, we can find total radiative properties of a plate,
(b) definition of gray bodies
(c) Kirchhoff's law
(d) greenhouse effect (or global warming effect) and other problems

15. References

1. James Hansen, et. al, *Earth's Energy Imbalance: Confirmation and Implications, Science*, vol. 308, pp. 1431–1435, 2005.

16. Exercise Problems

20-1 Consider a piece of gray-body glass situated in outer space, facing a blackbody heat source at T_source = 2000K, as shown in Fig. 20-11. The glass receives 70% of the energy emitted by the heat source. Also, $\rho = 0.3$ and $\tau = 0.6$. Find Ts.

20-2 In Eq. (10), if the surrounding air temperature is colder than Ts, then:
(a) We should modify the equation into

$$mc_v \frac{dT}{dt} = hA\,(-T_\infty + T) + A\,\alpha\,G - A\epsilon\sigma\,T^4$$

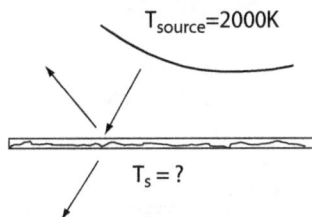

Fig. 20-11   A plate exposed to a heat source

(b) We should leave it unchanged.

(c) Both (a) and (b) are correct. The final solution will take care of the sign automatically.

(d) We should do (a) and make sure all T's are in the unit of K.

20-3 If an opaque gray-body plate is subject to a strong incoming radiation of 40,000 W/m^2, and if the plate initially is at 0C, then with m = 10 kg, ε = 0.6, and A = 1m^2, within the first 30 seconds, Ts increases with time

(a) linearly, (b) quadratically, (c) exponentially, (d) linearly then exponentially.

20-4 It is important for us to learn the basic skill to solve a time-dependent first-order one-variable differential equation other than using the explicit scheme. Consider a radiation-related cooling problem. After Ndm, the governing equation becomes

$$\frac{d\phi}{dt} = -c_1\phi^4$$, subject to $\phi(0)=1$, where ϕ could denote, for example, T_coffee (t)/T_coffee_i

Find ϕ (t) versus t and find at the 21st time step, with Δt = 5 and c1 = 0.1. Use the implicit method and the Newton-Raphson method.

20-5 The primary purpose of this problem is to demonstrate the validity of Kirchhoff's law. Consider an opaque plate whose data are given as:

mass of the plate = 10 kg; cv = 1000 J/kg-K; A = 1m^2;
Ts_initial = 1000K;

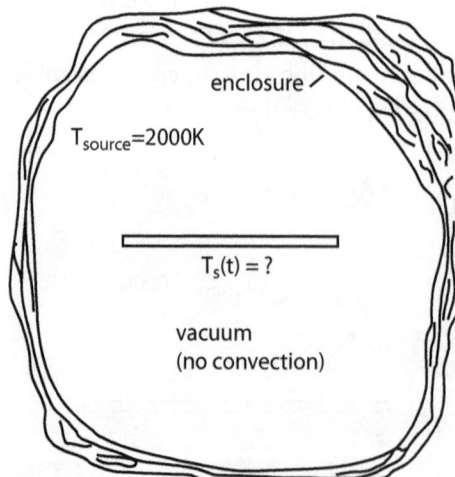

Fig. 20-12   A system consisting of an enclosure, a plate, and vacuum

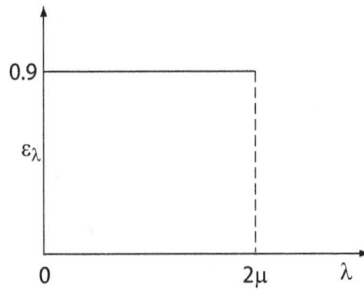

Fig. 20-13   Spectral emissivity distribution

The blackbody enclosure is at T_source = 2000K. Figure 20-12 shows the schematic system.

Take Δt = 0.2 sec; ntime = 500;
For simplicity, use the explicit method. The solution should be stable with such a small time step.

(a) The plate is gray with ε = 0.4; Plot Ts versus time between [0 100s].
(b) The plate is non-gray. The spectral emissivity distribution is shown in Fig. 20-13. Kirchhoff's law applies. Also plot Ts vs. time between [0 100s]. If our solution of Ts does not level off at 2000K, it is obviously incorrect.
(c) same as (b) except Ts_initial = 2800K.
(d) same as (b) except that we will pretend that Kirchhoff's law does not apply. What is the steady state temperature of the plate? (take ntime = 800) Does it make sense to us? We should immediately build a 2-T engine, generating free electricity, and becoming millionaires very quickly.

Lesson 21
Radiation (III)

$\mathbf{I}$n this lesson, the key topics to be studied are radiation exchanges inside a vacuum enclosure composed by not more than three surfaces, for the purpose of simplicity. These surfaces are all flat, infinitely deep perpendicular to the x-y plane, and can be either black or gray.

Nomenclature

F = view factor, the subscript 12 stands for "A1 to A2"

J = radiosity, defined as radiation leaving the surface, including emission and reflected radiation

ε_s = emissivity of the shield

1. View Factors (or Shape Factors, Configuration Factors)

Consider two surfaces situated near each other, as shown in Fig. 21-1. It is clear that only a fraction of all the radiation leaving surface A1 will arrive at A2. This fraction is related only to the geometry of the two-plate system. Prior to studying radiation exchanges between these two surfaces, we need to know how to calculate the value of such a fraction.

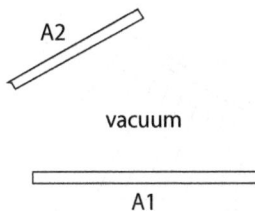

A2

vacuum

A1

Fig. 21-1   Two plates facing each other in vacuum

1-1 Definition of F_{12}

Let us define

F_{12} = (radiation leaving A1 and reaching A2) / (total radiation leaving A1).

Under this definition, obviously, $F_{12} \leq 1$.

1-2 Reciprocity Rule

It can be proved [1] that

$$A1\ F_{12} = A2\ F_{12} \tag{1}$$

In this textbook, we will simply assert that Eq. (1) does not contain any information regarding T1 and T2. Therefore, the direction of radiative fluxes inside the shaded area spanned by A1 and A2, shown in Fig. 21-2, is irrelevant. The energy leaving A1 that can possibly reach A2 can be represented by the shaded area, and vice versa. So, we conclude that Eq. (1) must be true.

1-3 Energy Conservation Rule

A triangular enclosure, consisting of A1, A2, and A3 as shown in Fig. 21-3, should bear the following relationship

$$F_{12} + F_{13} = 1 \tag{2}$$

We will concentrate on triangular enclosures, because we can readily extend our learning of 3 surfaces to 2, 4, 5, and more. Furthermore, let us try to avoid using subscripts i, j, and k. Our minds usually do not function efficiently when they see those abstract indices.

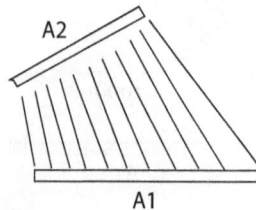

Fig. 21-2   Two plates face each other, and exchange radiation confined in the shaded area.

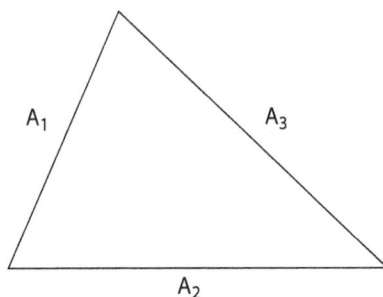

Fig. 21-3   An enclosure consisting of 3 flat surfaces

1-4 View Factor for a Triangle

The view factor of a triangular enclosure can be derived to be

$$F_{12} = (A1 + A2 - A3) / (2*A1).\tag{3}$$

See Problem 21-1. Values of view factors for other geometric systems can be found in [1].

2. Black Triangular Enclosures

Consider energy balance over blackbody surface 1 first. We have

$$q_1 = \text{out} - \text{in},$$

or

$$q_1 = A_1\sigma T_1^4 - A_2 F_{2-1}\,\sigma T_2^4 - A_3 F_{3-1}\,\sigma T_3^4 ,$$

or, using the reciprocity and dividing both sides by A1, we can further obtain

$$q_1'' = \sigma T_1^4 - F_{1-2}\,\sigma T_2^4 - F_{1-3}\,\sigma T_3^4 .\tag{4}$$

Energy balance over surfaces 2 and 3 can be conducted similarly. Consequently, there will be three energy-balance equations, and there are six unknowns: q_1'', q_2'', q_3'', T1, T2, and T3. Values of three quantities should be given in order for us to solve the problem.

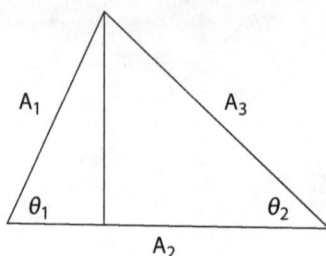

Fig. 21-4   Areas of three plates cannot be arbitrarity given.

Example 21-1

Prove that areas of A1, A2, and A3 in a triangle cannot be arbitrarily given (for example, 3, 4, and 8). In fact, any area of one surface must be smaller than the sum of other two areas.

Proof: In reference to Fig. 21-4, we have

A2 = A1*cos(θ1) + A3*cos(θ2).

But cos(θ1) < 1 and cos(θ2) < 1 must be true. Therefore,

A2 < A1 + A3 must be true. When we assign a problem, or are assigned a problem, we should watch out for this inequality.

Example 21-2

Consider a triangular enclosure consisting of A1 = 1, A2= 0.8, and A3 = 0.5043. All are blackbody surfaces.

(a) Given: T1=1000K, T2 = 1500K, and T3 = 2000K, which is the simplest case.

Find: q1, q2, and q3 such that the enclosure system is in thermal equilibrium. The solution obtained in (a) will be used as data input for cases (b) and (c).

(b) Given: q1, q2, and q3 (using values found in (a))
Find: T1, T2, and T3

(c) Given:T1, q2, and q3
Find: q1, T1, and T3

Sol: See the Matlab code below.

clc; clear % T1, T2, and T3 are all given

```
sig=5.67;
T1 = 10; T2 = 15; T3 = 20;
A1=1; A2 =.8; A3= 0.5043;
F12 = (A1 + A2 - A3)/(2*A1); F13 = 1-F12;
F21 = A1*F12/A2; F23 = 1-F21;
F31 = A1*F13/A3; F32 = 1-F31;
%
q1 = sig*T1^4 - F12*sig*T2^4 -F13*sig*T3^4
q2 = - F21*sig*T1^4 + sig*T2^4 -F23*sig*T3^4
q3 = - F31*sig*T1^4 - F32*sig*T2^4 + sig*T3^4
%
Qt = A1*q1 + A2*q2 + A3*q3 % check if Qt = 0;
%
%>>>>>>>>>>>>>>>> q1, q2, and q3 are all given (case (b))
q1s = q1/sig; q2s = q2/sig; q3s = q3/sig;
a(1,1)=1 ; a(1,2)= -F12; a(1,3)= -F13;
a(2,1)= -F21; a(2,2)=1; a(2,3)= -F23;
a(3,1)= -F31; a(3,2)= -F32; a(3,3)=1;
b(1)= q1s; b(2)= q2s; b(3)= q3s;
T4 = a\b' % singular matrix; problem ill-defined
%
%>>>>>>>>>>>>>>>> case (c)
sig=5.67; % If e-8 is dropped in sig, all T values are divided by 100.
A1=1; A2 =.8; A3=0.5043;
F12 = (A1 + A2 - A3)/(2*A1); F13 = 1-F12;
F21 = A1*F12/A2; F23 = 1-F21;
F31 = A1*F13/A3; F32 = 1-F31;
T1 = 10; q2s= 6.8568e4/sig; q3s =7.81e5/sig;
% if surface 3 is insulated, q3s = 0.
a(1,1)=1; a(1,2)=-F23; b(1)=q2s + F21*T1^4;
a(2,1)=-F32; a(2,2)=1; b(2)=q3s + F31*T1^4;
T4 =a\b';
T2=T4(1)^.25 % = 14.9997 ~ 15
T3=T4(2)^.25 % = 19.9999 ~ 20
```

3. Gray Triangular Enclosures

If an enclosure consists of non-black surfaces that are reflective, we need to exercise our wisdom to conduct the energy-balance analysis. If not, such an analysis can become painstakingly complicated. The wisdom of some brilliant scholars has taught us to introduce a quantity, called radiosity.

3-1 Definition of Radiosity, J

In reference to Fig. 21-5, the radiosity of surface $J1$, is defined as all the radiation leaving surface 1. It includes the radiation emitted by the surface itself, and the reflected portion of incoming radiation. Thus,

$$J_1 = \epsilon_1 \sigma T_1^4 + \rho G_1 \tag{5a}$$

where $G1$ is the incoming radiation impinging on surface 1. If surface 1 is gray, then Eq. (5a) can be rewritten as

$$J_1 = \epsilon_1 \sigma T_1^4 + (1 - \epsilon_1) G_1 \tag{5b}$$

Next, the energy balance over A_1 should be

$$q_1 = \text{out} - \text{in} = A_1 (J_1 - G_1). \tag{6}$$

A relationship between q1 and J1 thus can be obtained from eliminating G_1 between Eq. (5b) and Eq. (6), after algebra, as

$$q_1 = \frac{A_1 \varepsilon_1}{1 - \varepsilon_1} (\sigma T_1^4 - J_1) \tag{7}$$

In a triangular enclosure, Eq. (6) becomes

$$q_1'' = J_1 - F_{1-2}J_2 - F_{1-3}J_3. \tag{8a}$$

Similarly for surfaces 2 and 3, we have

$$q_2'' = J_2 - F_{2-1}J_1 - F_{2-3}J_3, \tag{8b}$$

and

$$q_3'' = J_3 - F_{3-1}J_1 - F_{3-2}J_2. \tag{8c}$$

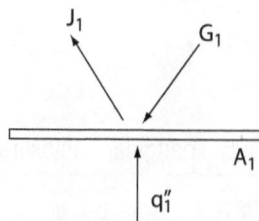

Fig. 21-5   Derivation of J1 and q1 relationship

If T_1, instead of q_1, is given, then Eq. (8a) can be modified into:

$$a(1,1)J_1 + a(1,2)J_2 + a(1,3)J_3 = \left(\frac{\varepsilon_1}{1-\varepsilon_1}\right)\sigma T_1^4 ,$$ (8d)

where a(1,1) = $1/(1 - \varepsilon_1)$, a(1,2) = $- F_{1\text{-}2}$, and a(1,3) = $- F_{1\text{-}3}$.

Example 21-3

Consider an equilateral triangular gray enclosure shown in Fig. 21-6 with $q_3 = 0$ and other relevant data given.

Find: T_3

```
clc; clear
sig=5.67e-8;
e1=.8; T1=1200;
e2=.4; T2=500;
e3=.8;
a= -.5*ones(3); % F12 = 0.5
a(1,1)=1/(1-e1); a(2,2)=1/(1-e2); a(3,3)=1;
b(1)=(e1/(1-e1))*sig*T1^4;
b(2)=(e2/(1-e2))*sig*T2^4;
b(3)=0;
qq=a\b'; % qq(3) is J3.
T3=(qq(3)/sig)^.25 % = 1102.2K
```

See Problem 21-2.

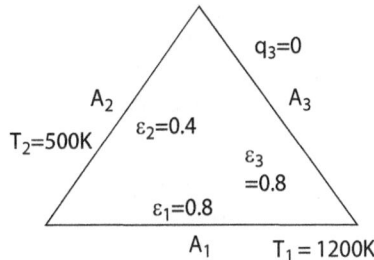

Fig. 21-6   The system schematic of a triangular enclosure

Let us now examine a special case. For two parallel surfaces that face each other shown in Fig. 21-7, Eqs. (8-a, b) can be further simplified into

$$q_1'' = J_1 - f J_2,\qquad\qquad(9a)$$

$$q_2'' = J_2 - f J_1,\qquad\qquad(9b)$$

where f is the view factor between the two plates. Together with Eq. (7) that is applied to both plates, we acquire two more equations:

$$q_1'' = \frac{\epsilon_1}{1-\epsilon_1}\left(\sigma T_1^4 - J_1\right)\qquad\qquad(9c)$$

and

$$q_2'' = \frac{\epsilon_2}{1-\epsilon_2}\left(\sigma T_2^4 - J_2\right).\qquad\qquad(9d)$$

There are four equations, and there are six unknowns: T_1, T_2, J_1, J_2, q_1'', and q_2''. When two of them are given, the problem can be solved.

Example 21-4

Given: Two parallel plates (A1 = A2 = $1 m^2$) are separated by a distance such that f = $F_{12} = F_{21} = 0.7$. The supplied heat fluxes and emissivities are given in the Matlab code below.

Find: T1 and T2
Sol:
clc; clear
sig=5.67e-8;

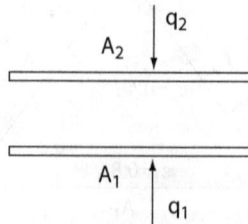

Fig. 21-7   **Two parallel plates facing each other**

```
q1=sig*(1000)^4; q2=sig*(1500)^4; f=0.7;
a(4,4)=0;
e1=.3; e2=.6; er1=e1/(1-e1); er2=e2/(1-e2);
a=[0 0 1 -f; 0 0 -f 1; er1 0 -er1 0; 0 er2 0 -er2];
b=[q1 q2 q1 q2];
qq = a\b';
T1 = (qq(1)/sig)^0.25 % = 1.832e3 K
T2 = (qq(2)/sig)^0.25 % = 1.957e3 K
E_escaped = (1-f)*(qq(3)+qq(4)) % = 3.437e5
E_input = q1+q2 % = 3.437e5
```

Near the end, we also check if the energy injected into the two-plate system is equal to the energy that has escaped from the system. It is observed that indeed they are equal, supporting the correctness of the analysis.

See Problem 21-3, where $A_1 \neq A_2$.

5. Radiation Shield

Finally, the last case we like to consider is radiation transfer between two parallel plates that are separated by an opaque shield, as shown in Fig. 21-8. Let us assume A1 = As = A2, otherwise the problem will become quite complicated. There are three plates. However, they do not form an enclosure. The shield is not supplied with any external energy, and its two sides are facing different heat sources.

Even though the system is different from triangular enclosures, the only challenging feature is the energy balance over the shield, rendering analyses to lie within the scope of our studies. It is recommended that we work on the case of the blackbody shield first. After having understood this case, then we advance to the case of the gray-body shield.

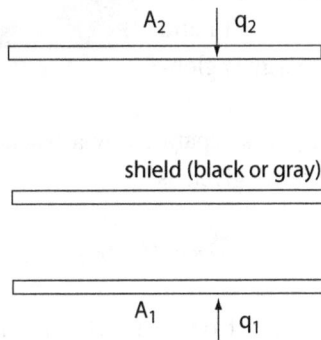

Fig. 21-8   Radiative transfer between two plates separated by a shield

See Problems 21-4 and 21-5.

6. Summary

In this lesson, we have studied:

(a) radiation exchanges among surfaces of a blackbody triangular enclosure,
(b) radiation exchanges among surfaces of a gray-body triangular enclosure,
(c) radiative transfer between two parallel plates separated by a gray shield.

7. References

1. Robert Siegel and John R. Howell, *Thermal Radiation Heat Transfer,* McGraw-Hill, 1972.

8. Exercise Problems

21-1 Prove that, for a triangular enclosure,

$$F_{12} = (A1 + A2 - A3) / (2*A1).$$

21-2 Revisit Example 21-3. If A3 is removed, what is the energy escaping from the system through this opening?

21-3 Revisit Example 21-4. Let $A1 = 1m^2$, $A2 = 3m^2$, and $F_{12} = 0.7$. Find T1 and T2, and check if the energy is globally conserved.

21-4 Consider two parallel plates separated by a blackbody shield as shown in the figure in Section 5. Data are given as

sig = 5.67e-8; q1=sig*(1000)^4; q2=sig*(1500)^4; F1s=0.7; F2s=0.4;

F1s stands for the view factor from plate A1 to shield.
Find Ts and check if the energy is globally conserved.

21-5 Repeat Problem 21-4, except now the shield is a gray body with ε = 0.3.

Credits

"Old But All with Colors," Source: http://www.flickr.com/photos/uaeincredible/205726358/. Cleared via Creative Commons Attribution 2.0 Generic license.

"Fatty Watching Himself on TV," Source: http://www.flickr.com/photos/46261842@N00/50257242. Cleared via Creative Commons Attribution 2.0 Generic license.

"Neely-Sieber House," Source: http://commons.wikimedia.org/wiki/File:Neely-Sieber_House,_front_with_driveway.jpg. Copyright in the Public Domain.

"Coca Cola C2," Source: http://commons.wikimedia.org/wiki/File:CocaCola_C2.jpg. Cleared via Creative Commons Attribution-Share Alike 3.0 Unported license.

"Birthday Candles," Source: http://commons.wikimedia.org/wiki/File:Birthday_candles.jpg. Cleared via Creative Commons Attribution-Share Alike 3.0 Unported license.

"Maternal Bond," Source: http://commons.wikimedia.org/wiki/File:MaternalBond.jpg. Cleared via Creative Commons Attribution-Share Alike 3.0 Unported license.

"Aeolipile Illustration," Source: http://en.wikipedia.org/wiki/File:Aeolipile_illustration.JPG. Copyright in the Public Domain.

"Lunar Eclipse 9-11-2003," Source: http://www.estelar.de/. Cleared via Creative Commons Attribution-Share Alike 3.0 Unported license.

"U.S. Navy Seabees with U.S. Naval Mobile Construction Battalion (NMCB) 15, Task Force Sierra, Prepare to Take Their Advancement Exams," Source: http://www.navy.mil/view_single.asp?id=55115. Copyright in the Public Domain.

"Windbeeches on the Schauinsland in Germany," Source: http://commons.wikimedia.org/wiki/File:Windbuchencom.jpg. Copyright in the Public Domain.

"SiS968," Source: http://www.flickr.com/photos/54299812@N06/5742856571. Cleared via Creative Commons Attribution-Share Alike 2.0 Generic license.

"Kangaroo Licking Itself to Cool," Source: http://en.wikipedia.org/wiki/File:Kangaroo_licking_itself_to_cool.jpg. Cleared via GFDL License found: http://www.gnu.org/copyleft/fdl.html

"ALSEP Apollo 14 RTG," Source: http://history.nasa.gov/alsj/a14/AS14-67-9366HR.jpg. Copyright in the Public Domain.

"Wonders of the Night Sky Map," Source: Popular Science Monthly, vol. 47. Copyright in the Public Domain.

"Spaghetti," Source: http://commons.wikimedia.org/wiki/File:Spaghetti.jpg. Cleared via Creative Commons Attribution-Share Alike 2.0 Generic license.

"African Bush Elephant," Source: http://commons.wikimedia.org/wiki/File:African_Bush_Elephant.jpg. Cleared via GFDL License found: http://www.gnu.org/copyleft/fdl.html

"Megafobia Oakwood Theme Park," Source: http://commons.wikimedia.org/wiki/File:Megafobia_Oakwood_Theme_Park.jpg. Cleared via Creative Commons Attribution ShareAlike 3.0 License.

"Café au lait," Source: http://www.flickr.com/photos/35756245@N00/18588671. Cleared via Creative Commons Attribution 2.0 Generic license.

"Swimming Breaststroke," Source: http://commons.wikimedia.org/wiki/File:Swimming.breaststroke.arp.750pix.jpg. Copyright in the Public Domain.

"Models at Michon Schur Fall 2007 Fasion Show, New York Fashion Week," Source: http://www.flickr.com/photos/52072922@N00/384711415/. Cleared via Creative Commons Attribution 2.0 Generic license.

"Հաւկիթի խաշել" Source: http://commons.wikimedia.org/wiki/File:%D5%80%D5%A1%D6%82%D5%AF%D5%AB%D5%A9%D5%AB_%D5%AD%D5%A1%D5%B7%D5%A5%D5%AC.JPG. Cleared via Creative Commons Attribution-Share Alike 3.0 Unported license.

"Vegan Patties with Potatoes and Salad," Source: http://www.flickr.com/photos/rusvaplauke/2298907166/. Cleared via Creative Commons Attribution 2.0 Generic license.

"World Trade Center, Aerial View March 2001," Source: http://commons.wikimedia.org/wiki/File:Wtc_arial_march2001.jpg. Cleared via Creative Commons Attribution-Share Alike 3.0 Unported license.

"Quinoa Soup," Source: http://commons.wikimedia.org/wiki/File:Quinoa_soup.jpg. Cleared via Creative Commons Attribution-Share Alike 3.0 Unported license.

"Croatia Airlines Airbus," Source: http://commons.wikimedia.org/wiki/File:Croatia.a320.arp.750pix.jpg. Copyright in the Public Domain.

"Kissing Prairie Dog," Souce: http://commons.wikimedia.org/wiki/File:Kissing_prairie_dog.jpg. Cleared via Creative Commons Attribution-Share Alike 3.0 Unported license.

"Suanpan and Soroban," Source: http://commons.wikimedia.org/wiki/File:Suanpan_and_soroban.jpg. Copyright in the Public Domain.

"Crystal Project Computer," Source: http://kde-look.org/usermanager/search.php?username=everaldo&action=contents. Cleared via GFDL License found: http://www.gnu.org/copyleft/fdl.html

"Jean-Claude Van Cauwenberghe," Source: http://commons.wikimedia.org/wiki/File:Van-cau-si-cravate.jpg. Cleared via Creative Commons Attribution-Share Alike 3.0 Unported license.

"Chickens at Tranquilles B&B," Source: http://commons.wikimedia.org/wiki/File:Chickens_at_Tranquilles_B%26B.JPG. Copyright in the Public Domain.

"Zydrunas Savickas Lifting Weight," Source: http://commons.wikimedia.org/wiki/File:Zydrunas_Savickas_lifting_weight.jpg. Cleared via Creative Commons Attribution-Share Alike 2.0 Generic license.

"Króliki," Source: http://commons.wikimedia.org/wiki/File:Kroliki001.jpg. Cleared via Creative Commons Attribution-Share Alike 3.0 Unported license.

"Critique of the Theory of Evolution Figure 017," Source: http://www.gutenberg.org/ebooks/30701. Copyright in the Public Domain.

"Capt. Robert E. Clark II Pins Meritorious Service Medal," Source: http://www.navy.mil/view_single.asp?id=74752. Copyright in the Public Domain.

"Home Sweet Home," Source: http://www.loc.gov/pictures/resource/pga.00261. Copyright in the Public Domain.

"OCP Musical Chairs," Source: http://commons.wikimedia.org/wiki/File:OCP_Musical_Chairs.jpg. Cleared via Creative Commons Attribution-Share Alike 3.0 Unported license.

"Slug," Source: http://commons.wikimedia.org/wiki/File:Slug_pic.jpg. Cleared via Creative Commons Attribution-Share Alike 3.0 Unported license.

"Archimedes Taking a Warm Bath," The Comic History of Rome. Copyright in the Public Domain.

"Take Five," Source: http://commons.wikimedia.org/wiki/File:Take_five.jpg. Cleared via Creative Commons Attribution-Share Alike 3.0 Unported license.

"Turbulent Water of Ken," Source: http://www.geograph.org.uk/photo/998757. Cleared via Creative Commons Attribution-Share Alike 2.0 Generic license. Attribution: Bob Peace.

"Lucky Bamboo," Source: http://commons.wikimedia.org/wiki/File:Lucky_bamboo.jpg. Cleared via Creative Commons Attribution-Share Alike 3.0 Unported license.

"Golf Ball Grass," Source: http://www.flickr.com/photos/34145688@N00/623791459. Cleared via Creative Commons Attribution 2.0 Generic license.

"U-tube Heat Exchanger," Source: http://commons.wikimedia.org/wiki/File:U-tube_heat_exchanger. PNG. Cleared via Creative Commons Attribution-Share Alike 3.0 Unported license.

"Redsilver Maine Coon Kittens," Source: http://commons.wikimedia.org/wiki/File:Redsilver_Maine_coon_Kittens.JPG. Cleared via Creative Commons Attribution-Share Alike 3.0 Unported license.

"Midsummer Bonfire Closeup," Source: http://commons.wikimedia.org/wiki/File:Midsummer_bon-fire_closeup.jpg. Cleared via Creative Commons Attribution-Share Alike 3.0 Unported license.

"Sun Pylon," Source: http://commons.wikimedia.org/wiki/File:Sun_Pylon.jpg. Cleared via Creative Commons Attribution-Share Alike 2.5 Generic license.

"Black Cat with Glowing Eyes," Source: http://commons.wikimedia.org/wiki/File:Black_cat_with_glow-ing_eyes.JPG. Cleared via Creative Commons Attribution-Share Alike 3.0 Unported license.

"A Small Cup of Coffee," Source: http://commons.wikimedia.org/wiki/File:A_small_cup_of_coffee. JPG. Cleared via Creative Commons Attribution-Share Alike 2.0 Generic license. Attribution: Julius Schorzman.

"Lit Jack-O'-Lantern Glowing Menacingly," Source: http://www.flickr.com/photos/7308682@N03/4041043478. Cleared via Creative Commons Attribution 2.0 Generic license.

"Rail Buckle," Source: http://www.volpe.dot.gov/sdd/buckling.html. Copyright in the Public Domain.

"Local Destinations," Source: http://www.geograph.org.uk/photo/486243. Cleared via Creative Commons Attribution-Share Alike 2.0 Generic license. Attribution: Graham Horn.

www.ingramcontent.com/pod-product-compliance
Lightning Source LLC
Chambersburg PA
CBHW080932220326
41598CB00034B/5758